Excel

Get the Results You Want!

SmartStudy 8

Science

Geoffrey Thickett
& Jim Stamell

PASCAL
PRESS

Reprinted 2015, 2018, 2020, 2021, 2022, 2023, 2024, 2025

ISBN 978 1 74125 421 1

Pascal Press
PO Box 250
Glebe NSW 2037
www.pascalpress.com.au

Publisher: Vivienne Joannou
Project editor: Leanne Poll
Edited by Leanne Poll
Indexed by Mary Coe
Typeset by Precision Typesetting (Barbara Nilsson)
Cover and page design by DiZign Pty Ltd
Photos by Dreamstime and iStockphoto
Printed by Vivar Printing/Green Giant Press

Students
All care has been taken in compiling this study guide, but please check with your teacher about the exact requirements of the course as these can change from year to year.

TABLE OF CONTENTS

Strand: Biological sciences 1

Cells 1
- Microscopes 1
- Cell structure 6
- Cell division and reproduction 12
- Disease 17

Body systems 24
- Multicellular organisms 24
- Digestive system 29
- Circulatory and respiratory systems 34
- Excretory and musculoskeletal systems 39
- Asexual and sexual reproduction 45
- Biotechnology 50

Strand: Chemical sciences 55

Elements and compounds 55
- Matter and changes of state 55
- Elements 61
- Compounds 66
- Simple chemical reactions 72

Strand: Earth and space sciences 77

Minerals and rocks 77
- Earth structure and minerals 77
- Weathering and erosion 82
- Sedimentary rocks and fossils 88
- Igneous and metamorphic rocks 93

Ores and environmental issues 98
- Metals and ores 98
- Extracting metals from common ores 104
- Maintaining our local environment 110
- Sustainability 116

Strand: Physical sciences 122

Energy 122
- Kinetic and potential energy 122
- Energy transformation 128
- Heat energy 133
- Energy efficiency 138

Skills 144

Investigations and problem solving 144
- Fair testing 144
- Working in the laboratory 150
- First- and second-hand investigations 156
- Experimental reports 161

Tips for the Sample Exam Papers 167

Sample Exam Papers 169
- Paper 1 169
- Paper 2 174

Answers 180
- Revision Tests 180
- Sample Exam Papers 207

Test & Exam Results 212

Feedback Checklist 213

Index 214

TAKE THESE REVISION STEPS TO SUCCESS!

Step 1 Quick Revision

- Check that you know the **key points** in each topic that you are studying by filling in the missing words in this section.
- Once you have completed the revision, **mark your work** by looking at the answers at the bottom of the page. This is **instant feedback** for you.

Step 2 Revision Summaries

- If you got any of the answers wrong in the **Quick Revision** section you can revise by reading the corresponding point in the **Revision Summaries** section. The numbered points in the **Quick Revision** section correspond to the numbered points in the **Revision Summaries** section.
- If you got all the answers right in the **Quick Revision** section then just skim over this section to be 100% sure you know the material.
- The **checklist** at the end of this section will help you **double-check** the essential information that you should know from this section.

Step 3 Revision Test

- These are test-style questions that you should be able to answer in order to prepare for the class test or common test. In this section you **recall** and **apply your knowledge**. Make sure you have **fully revised** your work in the **Revision Summaries** section.
- **Hints** for some questions are provided in case you need extra help with them. They are found at the bottom of the page.
- **Marks** are allocated for each question. Always check the marks and use them as a guide for **how much to write** in your answer.
- Time yourself—**check** how much **time** you have got to complete the test. Also look at the **total marks** of the test to calculate approximately how much time you should spend on each question. For example, if there are twenty marks in total and twenty minutes have been allocated for completion of the test, then spend about one minute on a one-mark question and six minutes on a six-mark question. If you cannot complete all the questions within the suggested time, you may need to revise the topic.
- Fill in the **Your Feedback** panel once you have marked your work in order to calculate your percentage mark. Then complete the **Test & Exam Results** page to keep a running total of all your test marks.

Step 4 Check Your Answers

- **Complete answers** to the **Revision Test** and **Sample Exam Paper** questions are also found at the back of the book. When you have finished a test, turn to this section to check the answers. Make a note of any questions that you got wrong.
- The **ticks** that appear in the answers indicate those parts of the answer which receive marks. Therefore, even if your answer is incomplete, you may be entitled to some marks for what you have written. Compare the answer to your own answer to find out whether you are entitled to any marks for the question.
- If you still **cannot understand** how the correct answer was obtained, revise that part of the topic and, if necessary, refer to your class notes or ask your teacher for assistance. It is important to learn from your mistakes. Remember: the questions in this book are very similar to the ones you will get in your examination. Using a pencil, write down the marks you get for each question, then add them up to get a total.
- Remember that there is an ***Excel** Year 8 Science Study Guide* available to help you revise further.

Step 5 Test & Exam Results

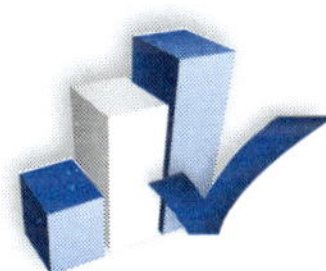

- Go to the **Test & Exam Results** section on page 212 to record your test score as a percentage. When you have completed all topics you will be able to determine your areas of weakness and your areas of strength.
- It is important that you know which **areas need further work**—to 'know what you don't know'. The more you prepare for the topic tests, the more successful you will be and the more you will remember when you sit your end-of-term/semester/year test or exam.

Step 6 Sample Exam Papers

- There are two Sample Exam Papers in this book. Sample Exam Paper 1 is similar to a **half-yearly test** and covers the Biological Sciences and Chemical Sciences strands. Sample Exam Paper 2 is similar to an **end-of-year test** that covers the entire contents of the book.
- Before attempting the **Sample Exam Papers**, make sure that you have completed all of the **Revision Tests** and have worked through the solutions to all the questions you answered incorrectly.
- Set aside the **time allowed** for the paper and complete it under **exam conditions**—no sneaking a look at your notes or textbook! That way you will be better prepared for your final exam.
- **Complete answers** to the Sample Exam Papers are provided at the end of the book.
- **Mark** your test to see how well you have done. Use the **Test & Exam Results** page to determine your areas of weakness and your areas of strength.

HOW TO USE THIS BOOK TO STUDY FOR A CLASS TEST, HALF-YEARLY OR END-OF-YEAR EXAM

Depending on your teacher or school, you will be given a variety of tests and exams each year. There may be a single-topic test, a test that covers a number of topics, a semester test or exam, or even a half-yearly or yearly exam.

Step 1 **Find out** which topics will be covered in the class test.

- To do this, look at your class workbook/textbook, laptop/tablet or online study program.
- Ask your teacher if you are still not sure what topics they are. For example, your class test may be on Cells.

Step 2 **Match** the topics that your test is on to the topics in this book.

- For example, the first four units in this book cover Cells.

Step 3 **Use** this book to study the topics being tested.

- For example, the first four units in this book contain the topics you will study for your class test!

Note:

- When you are using the book to study for a **half-yearly** test, follow the same steps as above—the only difference being that will be you will have more topics to revise of course!
- When you are using this book to study for an **end-of-year** test, you will more than likely need to study the whole book!

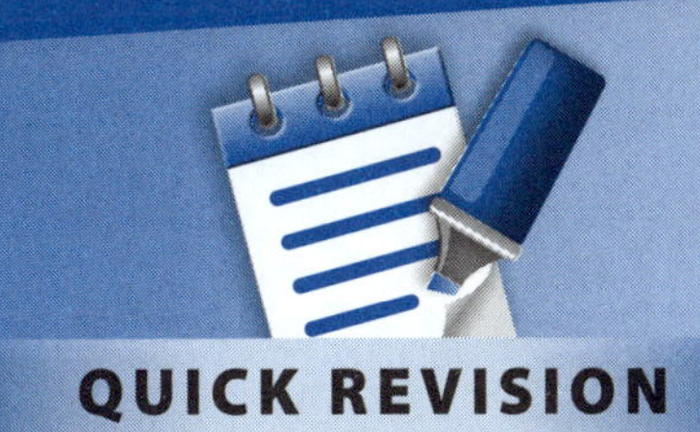

MICROSCOPES

Cells

QUICK REVISION

1 It was not until the ________ century that telescopes and microscopes were constructed using glass lenses. The first two-lens microscope was constructed by Robert ________ and this microscope had only a very small magnification. Hooke examined thin slices of materials such as ________ and found that this tissue contained small honeycomb-like structures. He called these compartments cells and this term is used today for the basic building ________ of all living things. As lens grinding technology improved, more ________ microscopes were developed. Antoine van Leeuwenhoek developed a ________-lens microscope that was nine times more powerful than Hooke's microscope. It soon became apparent to van Leeuwenhoek that there was a huge microscopic world in nature. Leeuwenhoek discovered that many organisms consist only of a ________ cell.

2 Microscopes are used today to investigate the structure of materials. The monocular microscope can be used in a school science laboratory to investigate various biological ________ such as the skin of an onion or a section of a muscle tissue. The tissue sample is placed in a drop of water on a glass ________ and then covered with a cover slip. The nosepiece of the microscope is rotated to select the ________ power objective lens. While looking through the ________, the lamp and the angle of the ________ are adjusted to get a bright circle of light in the field of view. The light intensity is adjusted with the ________ lever. The slide to be examined is placed on the stage. Looking from the side, the tube is lowered until the ________ lens is just above the slide. While looking through the eyepiece, the ________ focus knob is rotated to wind the tube up until the specimen is in focus. The position of the slide is adjusted and the ________ focus knob is rotated to complete the focusing. Binocular microscopes are also used in a school laboratory. These microscopes are used to examine ________ specimens such as crystals, seeds, flowers and small insects. The specimen is placed on the illuminated ________ and the low power objective lens is selected. There are ________ eyepiece lenses to allow a three-dimensional view by using both eyes. A higher power objective lens is selected if additional magnification is required. These microscopes also allow scientists to accurately dissect tissue and are used in dentistry and surgery.

3 Microscopes are used to produce a magnified image. Different-shaped lenses produce different degrees of ________. A lens that is marked 100× is one that produces an image that is 100 times ________ than the object. A microscope uses several lenses together to increase the magnification of the specimen. The total magnification for a two-lens system can be found by ________ the magnifications of both lenses together. Thus, a 10× eyepiece lens used in conjunction with a 40× objective lens produces a ________ magnification of 400×. When using low or high power, the diameter of the ________ of view can be determined by examining a clear millimetre grid. If five grid lines can be seen across the field then the field diameter is ________ mm. At higher magnifications, smaller scales must be viewed. Digital or video cameras can be attached to the eyepiece lens to observe and record images of the specimen. This allows many people to observe the ________ on a monitor. Microscopes that use light can only ________ specimens by about 2000 times. Electron microscopes are used to see finer details. These use beams of ________ instead of light rays and they can see deep into the structure of a cell. Such microscopes can magnify about 10 million times.

Answers **1** 16th; Hooke; cork; blocks; powerful; single; single **2** tissues; slide; lowest; eyepiece; mirror; iris; objective; coarse; fine; large; stage; two **3** magnification; larger; multiplying; total; field; 5; image; magnify; electrons

MICROSCOPES

Cells

1 At the end of the 16th century, the **magnifying glass** or lens became the first scientific instrument to examine the structure of living matter. In 1665 Robert Hooke invented the first two-lens **microscope** which could magnify about 30 times. He examined a thin slice of cork with his microscope and found it was composed of small compartments that he called **cells**. Improvements in lens manufacture soon followed and in 1676 Antoine **van Leeuwenhoek** developed a single-lens microscope that could magnify 270 times. Many cellular life forms were discovered with this microscope including single-celled organisms called **protozoa**.

2 Two common microscopes that are used in the school laboratory are called monocular and binocular microscopes. The following diagram shows the structure of a typical **monocular microscope**. The lamp provides light to illuminate the specimen. The specimen is viewed with light transmitted through it and so the specimen must be very thin.

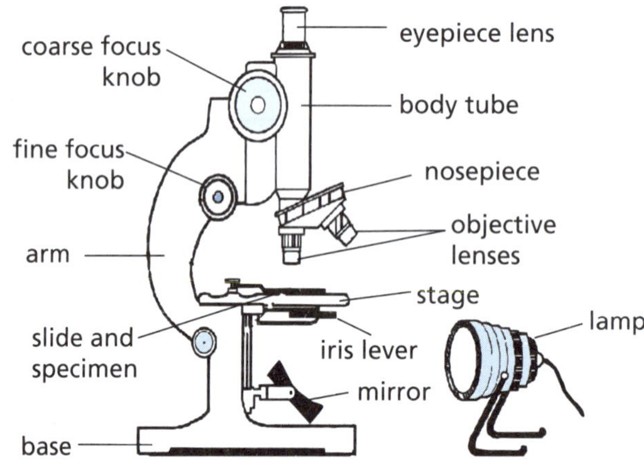

The following table lists the function of each major component of the monocular microscope.

Component	Function
eyepiece lens	low-power lens system used by the eye to examine the specimen
objective lens	high-power lens used to magnify the specimen
nosepiece	rotates to allow different objective lenses to be selected
coarse focus	knob used to raise/lower the body tube to bring the image into approximate focus
fine focus	knob used to bring the image into sharp focus
stage	the platform on which the slide and specimen are placed
iris	an adjustable diaphragm used to alter the amount of light on the specimen
slide and specimen	the specimen is placed on a glass slide which is placed on the stage
mirror	reflects light (from a lamp or the Sun) onto the specimen

The basic components of a **binocular microscope** are shown in the diagram below. The two eyepiece lenses allow you to view the specimen with both eyes. This type of microscope is used for examining large specimens (e.g. flowers, minerals and insects) where high magnification is not required. The specimen is placed on the stage and illuminated by a lamp. The reflected light passes through the lens system into your eye. Usually there are two different objective lenses of different magnification which can be slid into place.

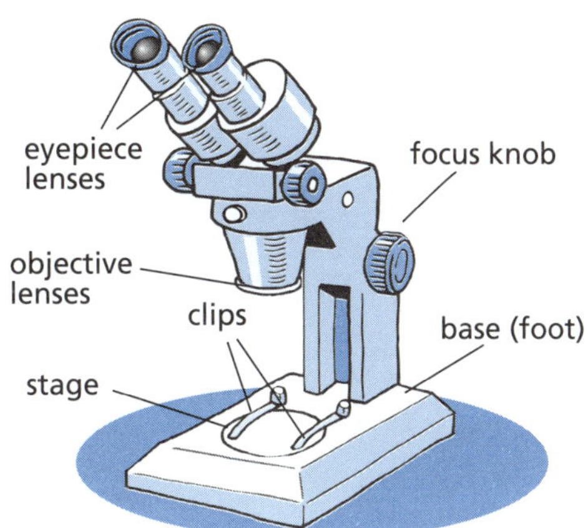

3 The lens system of a microscope allows the specimen to be magnified so its structure can be more easily seen. Each lens has a number engraved on the side that tells you its **magnification**. The typical eyepiece lens magnifies the image by 10 times (10×). Common objective lenses magnify the image by 10×, 20× and 40×.

You can calculate the overall magnification by multiplying the magnification of the eyepiece and objective lenses.

total magnification = eyepiece magnification × objective magnification

For example, if the eyepiece is 10× and the objective is 20× then total magnification = 10 × 20 = 200×

The total magnification can be used to determine the dimensions of the objects being viewed. The diameter of the field of view must first be determined. In order to do this, a millimetre scale (grid) is viewed at a known magnification. The following diagram shows a millimetre grid on low power (100×). In this example the diameter of the field of view is about 3 mm.

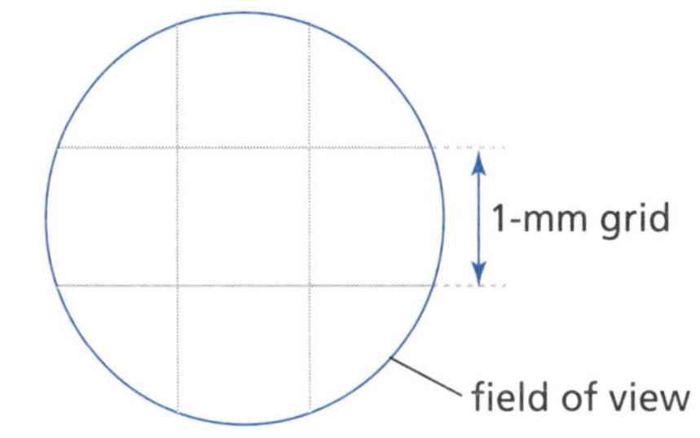

In many laboratories, digital or video cameras are attached to the eyepiece lens of a microscope. This allows us to capture and record magnified images and videos on computers for further analysis. The best light microscopes can achieve magnifications of 2000×. When much greater magnification is required, **electron microscopes** can be used. Electron microscopes use beams of electrons instead of light to create magnified images of a specimen. They can achieve magnifications of 10 000 000×. This allows scientists to view objects as small as 50 pm (picometres; 1 pm = 1 trillionth of a metre = 10^{-12} m).

Checklist

Can you:

1 *Describe the historical development of light microscopes?* ☐
2 *Describe the structure of monocular and binocular microscopes?* ☐
3 *Explain how to calculate the magnification of an image produced by a light microscope?* ☐

MICROSCOPES
Cells

REVISION TEST

1 The following diagram shows a millimetre grid viewed under low power (magnification: eyepiece lens = 10×; low power objective = 10×). Some small round seeds are also shown at the same magnification.

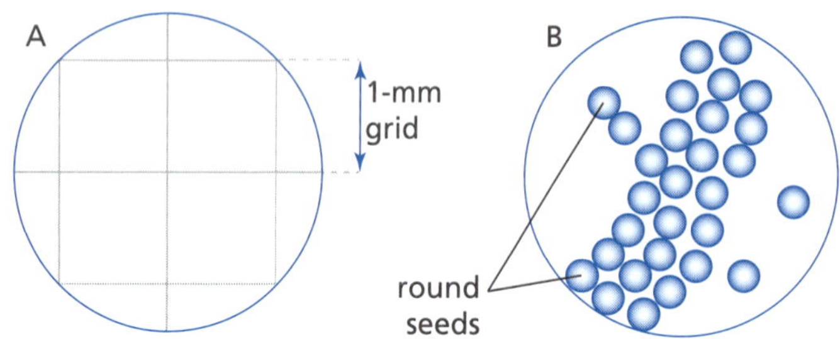

a What is the total magnification of the low power field? (1 mark)

b What is the diameter of the low power field in the following units?

  i millimetres (1 mark)

  ii micrometres *Hint 1* (1 mark)

c What is the average diameter (in micrometres) of one seed as viewed in the low power field? (1 mark)

d A higher-powered objective lens (20×) is now selected to examine the same seeds.

  i What will be the total magnification now? (1 mark)

  ii What will be the new field diameter (in micrometres)? (1 mark)

2 Identify three differences between monocular and binocular microscopes. (3 marks)

3 The following diagram shows different microscopic organisms that live in the water. The magnification is shown next to each organism.

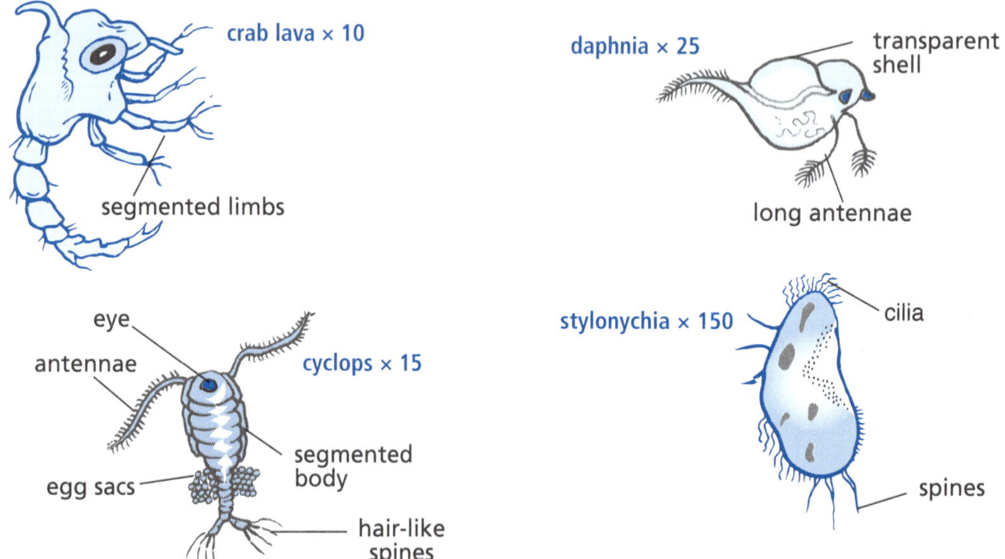

a Which organism is the largest? *Hint 2* (1 mark)

b Which organism is the smallest? (1 mark)

4 The following table lists the components of a monocular microscope and their function. The items, however, are jumbled. Match the letters in the first column to the numbers in the second column. (5 marks)

Component		Function	
A	fine focus	1	reflects light (from a lamp or the Sun) onto the specimen
B	objective lens	2	knob used to bring the image into sharp focus
C	nosepiece	3	high power lens used to magnify the specimen
D	mirror	4	an adjustable diaphragm used to alter the amount of light on the specimen
E	iris	5	rotates to allow different objective lenses to be selected

5 The following procedure is used to examine a specimen using a binocular microscope. The steps of the procedure are jumbled. Reorder the numbered steps in the correct sequence. (2 marks)

1 Select the lowest power objective lens.

2 Place the specimen on a tile or glass dish and shine the lamp to illuminate the specimen on the stage.

3 Lower the tube by looking from the side and then use the focus knob to raise the tube and focus while looking through the eyepiece lenses.

4 Place the binocular microscope carefully on the bench and arrange a lamp so its light will illuminate the stage.

6 When using a monocular microscope, the correct procedure is to lower the tube while looking from the side and then raising the tube to achieve a focus. The following photo shows a glass specimen slide being examined microscopically. Why is it incorrect to focus down onto the specimen rather than focus upwards? *Hint 3* (1 mark)

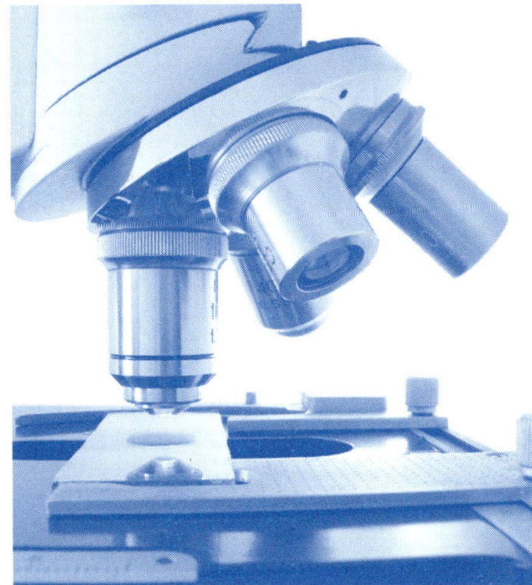

Hint 1: There are one thousand micrometres in one millimetre.

Hint 2: A large magnification indicates that the organism is quite small.

Hint 3: The lenses are expensive and should not be damaged.

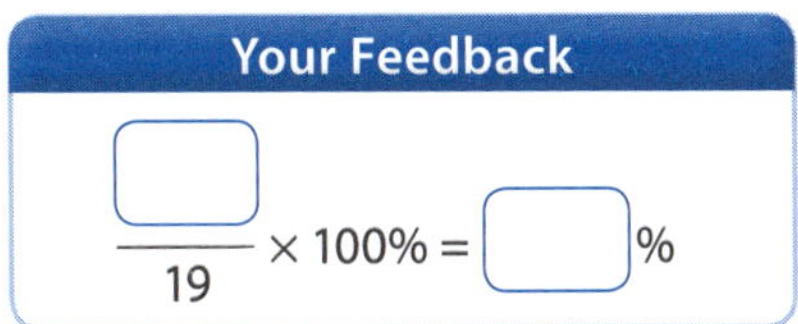

CELL STRUCTURE

Cells

1 Throughout the 18th and 19th centuries various lens grinders improved on the quality of lenses that were used in microscopes and consequently the degree of __________ increased. Biologists used these improved microscopes to study the structures of tissues from a wide variety of animals, plants, fungi and unicellular organisms. It became apparent that all living tissue is composed of __________ and that new cells form from __________ cells by a process of cell division and growth. The contents of the cell are called __________ and a membrane surrounds this watery material. The protoplasm consists of a __________ surrounded by a watery cytoplasm. Inside the cell are a variety of smaller structures called __________ within the cytoplasm. The nucleus controls the activities of the cell. The cell membrane is selectively permeable and allows materials to __________ in and out of the cell.

All cells need nutrients and __________ to survive and these materials diffuse into the cell whereas wastes diffuse in the __________ direction. The cellular activities require energy and this energy is produced when nutrients such as glucose are converted to carbon dioxide and water by organelles called __________. Some of the cells in a green plant are responsible for photosynthesis. In this process a plant cell converts carbon dioxide and __________ into oxygen and glucose. The energy required for this process comes from the __________. Solar energy is absorbed by organelles called __________ which contain the green pigment called __________.

2 Inside your body and in the bodies of all animals there are many different types of cells and each type of cell has a specific __________. Our bodies require oxygen and this is transported to all our cells via __________ blood cells. Some microbes cause disease, and one of the cells that protect us from microbe attack is the __________ blood cell. Information from our brain is transferred to various organs of our body via __________ cells. Reproduction occurs when an egg cell is __________ by a sperm cell. The following diagram shows a human sperm at the surface of a human egg cell. Our muscle cells can __________ and expand, and these forces allow us to move our bodies.

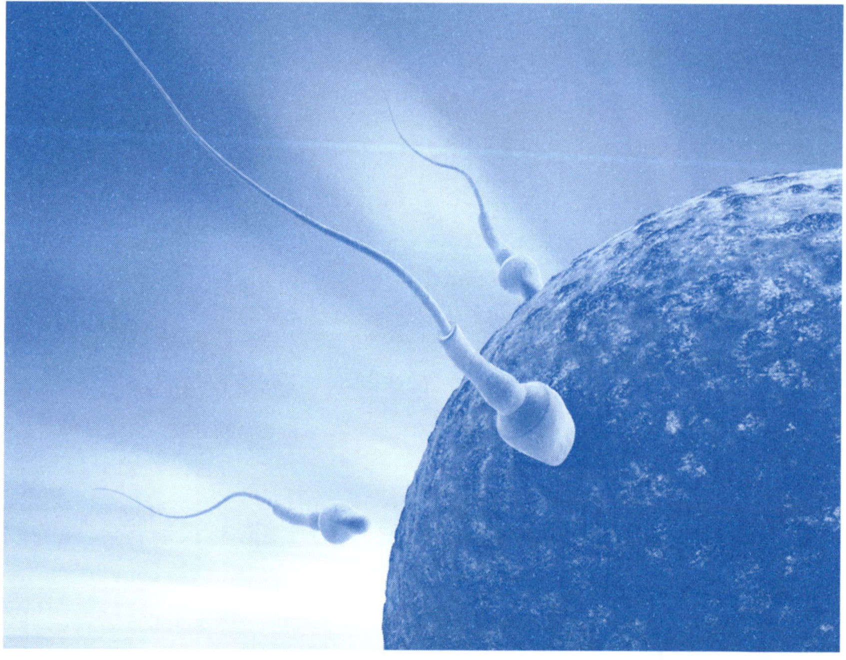

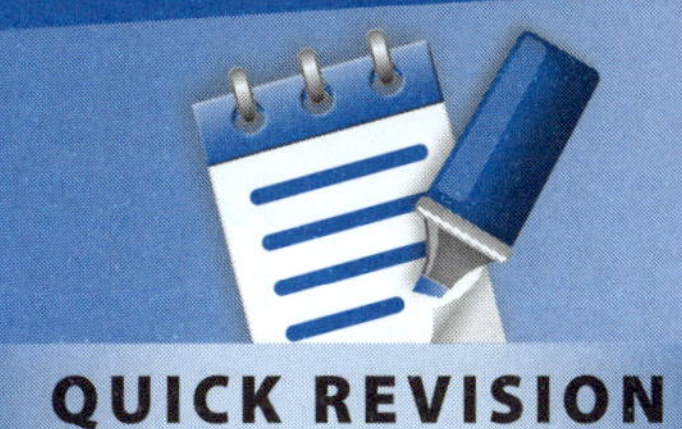

3 Plants cells are similar to animal cells but they also possess a cell __________ which surrounds the whole cell. This wall is composed of __________ which is the major constituent of wood. The cell wall provides support for the cell and allows trees to grow tall. In leaves and some stems there are cells that contain __________ called chloroplasts. These chloroplasts contain a pigment called __________ which absorbs sunlight. This solar energy is used to power a chemical process called __________ in which water and carbon dioxide are converted to __________ and oxygen. The oxygen is used by the plant and the excess is released into the atmosphere for animals to breathe. The glucose is transported to all parts of the plant in tubes formed from cells called __________.

A model of the structure of a glucose molecule is shown in the following diagram. Glucose is made of six carbon atoms, twelve hydrogen atoms and six oxygen atoms.

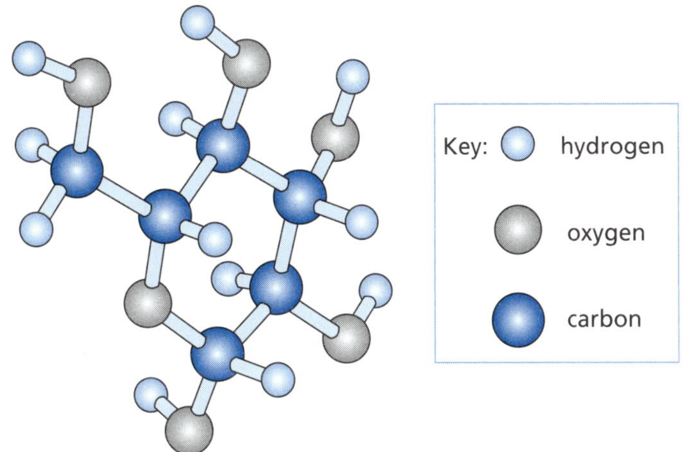

Most plant cells have large __________ which store water and some nutrients and waste. When the vacuole is filled with __________, the plant cannot wilt on hot days. If water is lost then the plant will wilt. The water that plants require comes from the soil. It is absorbed through root hairs and it then moves up the plant in tubes made of cells called __________. Leaves have __________ (mainly on their lower surfaces) called stomata. These stomata open and close to allow gases as well as water vapour to be exchanged with the atmosphere. __________ cells are responsible for opening and closing the stomata (also known as stomates).

4 In the five-kingdom classification of living things, one kingdom is called protista and this kingdom mainly contains __________ organisms. It also includes some organisms that have very simple cells joined together forming tissues. A protist called euglena is able to photosynthesise as it contains __________. It is also able to move about as it has a whip-like __________ called a flagellum. Some protists have two flagella. The protist called paramecium has an opening in its cell called an oral groove through which __________ may enter. The paramecium can move about with the aid of fine hair-like structures called __________ which move and create propulsion. Protists are commonly found in bodies of water including soil water.

Answers **1** magnification; cells; old; protoplasm; nucleus; organelles; diffuse; oxygen; reverse (opposite); mitochondria; water; Sun; chloroplasts; chlorophyll **2** function; red; white; nerve; fertilised; contract **3** wall; cellulose; organelles; chlorophyll; photosynthesis; glucose; phloem; vacuoles; water; xylem; pores; guard **4** unicellular; chloroplasts (chlorophyll); tail; food; cilia

CELL STRUCTURE

Cells

1 By 1838 continued improvements in microscope design and the microscopic examination of a wide variety of plant and animal tissue allowed two German scientists (Matthias Schleiden and Theodor Schwann) to propose the **cell theory** of living matter. The following are the major points of the modern cell theory.

- All living things are made up of one or more cells.
- Cells are the smallest units of living things.
- All cells come from other pre-existing living cells.

All living cells are composed of watery **protoplasm** surrounded by a **cell membrane**. Within the protoplasm are smaller structures called **organelles**. The protoplasm is divided into two parts called the **nucleus** and the **cytoplasm**. The organelles are located in the cytoplasm.

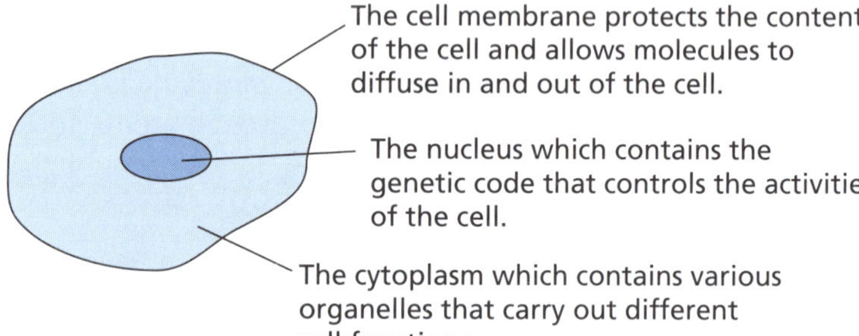

The cell membrane allows various nutrient molecules to enter the cell and to allow waste materials to leave the cell. This two-way movement of molecules or ions is called **diffusion**. Diffusion occurs when the concentration of a substance is greater on one side of a membrane than the other. As wastes build up in a cell they diffuse out of the cell into a region of lower concentration. Diffusion always occurs from a region of high concentration to a region of lower concentration.

Cells contain a number of different organelles. The following lists three common organelles and their functions.

Organelle	Function
mitochondrion	organelle where glucose is converted to usable energy by a process called cellular respiration
vacuole	organelle where water, food and waste substances are stored; plants have large vacuoles but they are often tiny or non-existent in animal cells
chloroplast	organelle found in green plants only; contains the green chemical called chlorophyll which absorbs sunlight; site of photosynthesis where solar energy is converted into chemical energy

2 There are various types of cells in animals. They have different shapes and functions. The following table lists some examples of cells present in animals and their functions. The diagram shows some examples of animal cells.

Cell type	Function
red blood cell	transports oxygen through the blood stream
white blood cell	attacks invading microbes
nerve cell	carries electrical communications in the nervous system
epidermal cell	cells that cover the surface of an organ or organism
egg cell	reproductive cell in females
sperm cell	reproductive cell in males
smooth muscle cell	allows muscles to work by contracting and expanding

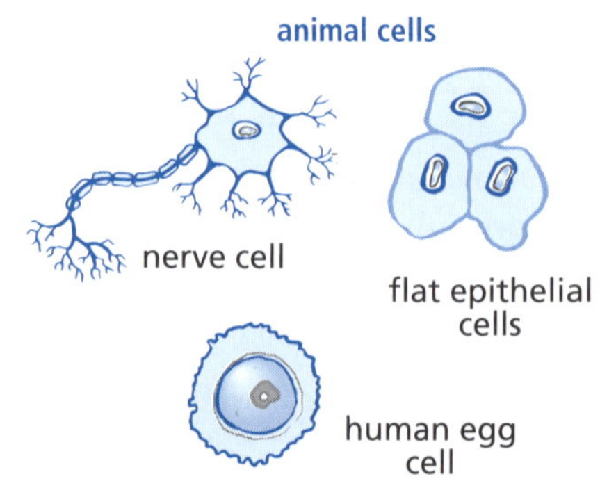

CELL STRUCTURE

Cells

3 There is a wide variety of cells in plants. Unlike animal cells, plant cells also contain a cellulose cell wall which provides strengthening and support for the plant. The following diagram shows that the **cell wall** is outside the **cell membrane**.

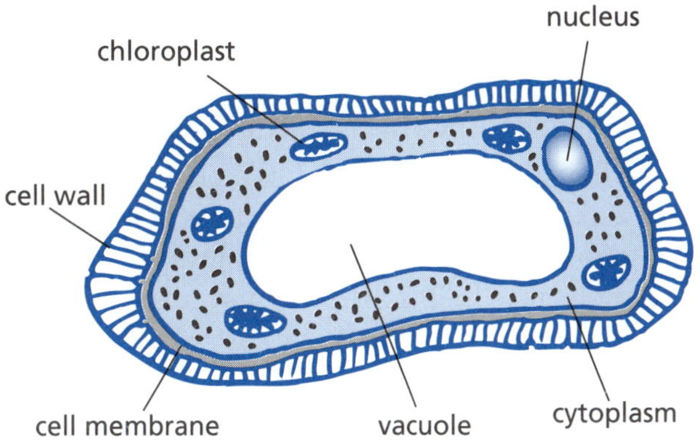

nucleus
chloroplast
cell wall
cell membrane
vacuole
cytoplasm

The **vacuoles** in plant cells are also quite large. Plants transport water throughout the plant in xylem cells. Phloem cells are used to transport nutrient solutions. Guard cells allow the leaf pores (stomata) to open and close to allow **gas exchange** with the atmosphere. Palisade cells in a leaf contain chloroplasts which are engaged in photosynthesis. The diagram shows a variety of plant cells.

plant cells

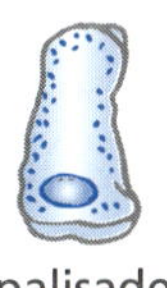
palisade cell

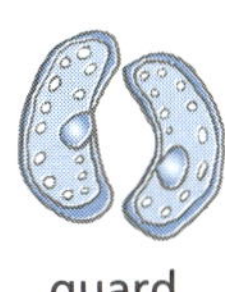
guard cells

phloem cell

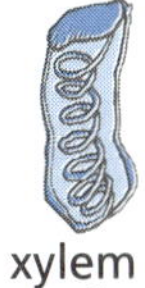
xylem cell

For many years fungi were classified in the same kingdom as plants. Today they are classified in their own kingdom. Two important reasons for classifying fungi in a separate kingdom is that they do not photosynthesise as they have no **chloroplasts** and their cell wall are made from a material called chitin rather than cellulose.

4 The kingdom protista contains many organisms that consist only of one single cell. They are called unicellular organisms. Some of these unicellular organisms may photosynthesise if the cell contains chloroplasts. Those that do not photosynthesise obtain their nutrition by nutrient absorption via the cell membrane or small openings (oral grooves) through which food may enter the cell. Some protists have whip-like tails called **flagella** in order to move about. Others use tiny hairs called **cilia** to assist movement. Two examples are shown here.

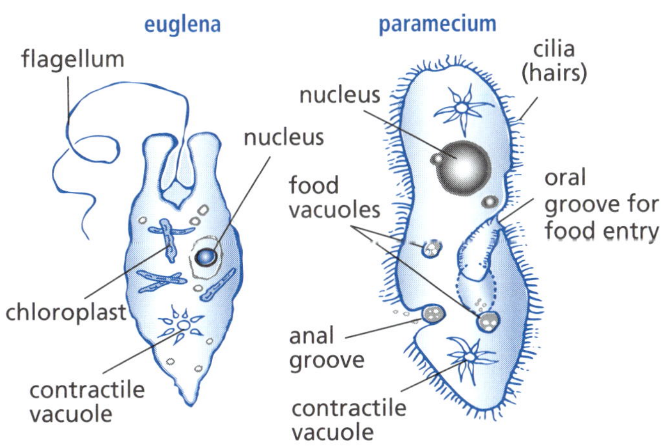

Checklist

Can you:

1 *State the major points of the cell theory and identify the function of the major organelles in a cell?* ☐

2 *State the function of common types of animal cell?* ☐

3 *State the differences between plant and animal cells and state the function of some common types of plant cells?* ☐

4 *Name some examples of unicellular organisms and name the kingdom to which they belong?* ☐

CELL STRUCTURE

Cells

REVISION TEST

1 Read the following text and answer the questions that follow.

> Membranes surround all living cells. Within the cells, membranes also surround many organelles such as the nucleus, the mitochondrion and the chloroplast.
>
> Membranes are not just like plastic bags which hold the cytoplasm and prevent it from leaking out. The basic structure of a membrane is a double layer of molecules including phospholipids, as well as cholesterol, proteins and carbohydrates. The surface carbohydrate molecules help cells recognise foreign materials. The cholesterol molecules help stabilise the membranes in animal cells whereas in plant cells the membrane is supported by a rigid cellulose cell wall.
>
> Membranes have a wide variety of functions.
>
> **1** Membranes allow cells to exchange substances with their external environment.
>
> **2** Membranes create specialised zones inside the cell.
>
> **3** Membranes restrict the movement of various substances inside the cell.
>
> **4** Membrane surfaces help cells to recognise foreign bodies.
>
> Membranes are not static structures. Fatty molecules in the membrane can move within the membrane layers. Proteins can provide channels for the passive diffusion of certain water-soluble molecules across the membrane.

- a Name the molecules that form a double layer in the cell membrane. (1 mark)
- b What is the purpose of the surface carbohydrates on a cell membrane? (1 mark)
- c State the function of protein channels in the cell membrane. (1 mark)
- d Which of the four listed functions is related to the diffusion of nutrients into a cell and the diffusion of wastes out of a cell? (1 mark)
- e Identify molecules or structures that help to support and stabilise cell membranes in the following.
  - i animal cells (1 mark)
  - ii plant cells (1 mark)

2 Name the cells that carry out each of the listed functions.

- a form water transport tubes in plants (1 mark)
- b carry electrical messages from one cell to another in animals (1 mark)
- c open and close the pores in a green leaf to exchange gases (1 mark)
- d fight against invading disease organisms in humans (1 mark)

3 Name the cellular structure or organelle that has each of the following functions.

- a the control centre of the cell (1 mark)
- b converts nutrients into usable energy (1 mark)
- c absorbs solar energy for photosynthesis (1 mark)
- d provides strength and support for the cell in a plant (1 mark)

4 State two reasons why fungi are no longer classified in the plant kingdom. *Hint 1* (2 marks)

5 The following diagram shows a protist called an amoeba and the way it feeds.

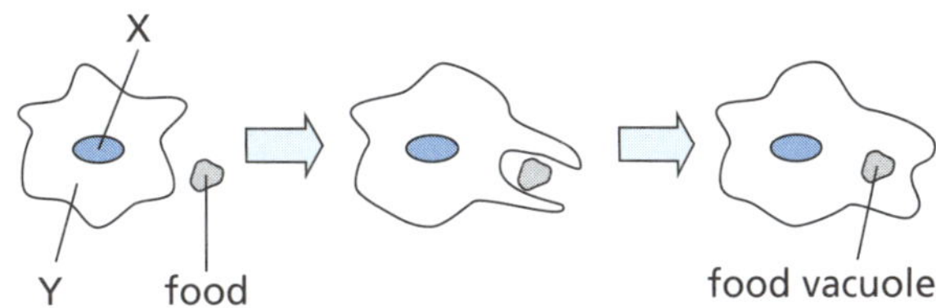

a Name the cell components labelled X and Y. *Hint 2* (2 marks)

b Define the term *protist*. (1 mark)

c Use the diagram to explain how the amoeba feeds. (2 marks)

6 A thin section of a green leaf was examined using a microscope. The following diagram shows a cross-section of this leaf.

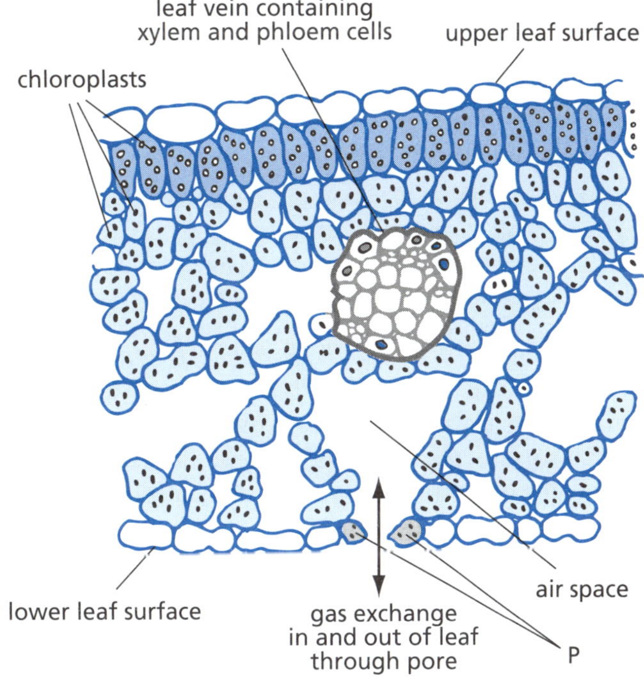

a A pore is visible on the lower surface. What is this pore called? (1 mark)

b Gases are exchanged in and out of the leaf through this pore.

  i Identify two gases that move outwards through this pore. (2 marks)

  ii Identify one gas that moves into the leaf through this pore. (1 mark)

c Cells (P) allow the pore to open and close. Name these cells. (1 mark)

d Suggest two roles for the leaf vein. *Hint 3* (2 marks)

Hint 1: Do fungi have chloroplasts?
Hint 2: X controls the cell's activities.
Hint 3: What do xylem cells do and what do phloem cells do?

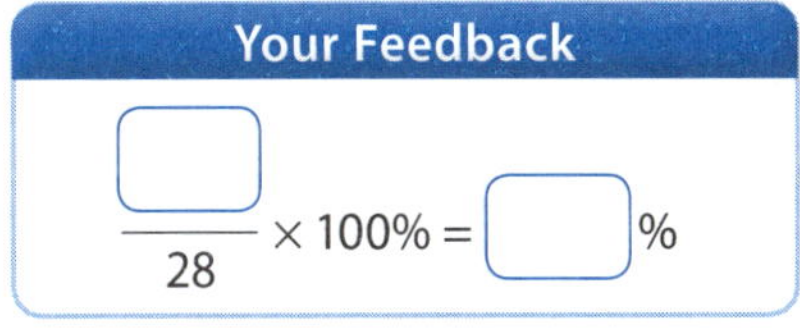

PAGE 180
PAGE 212

CELL DIVISION AND REPRODUCTION

Cells

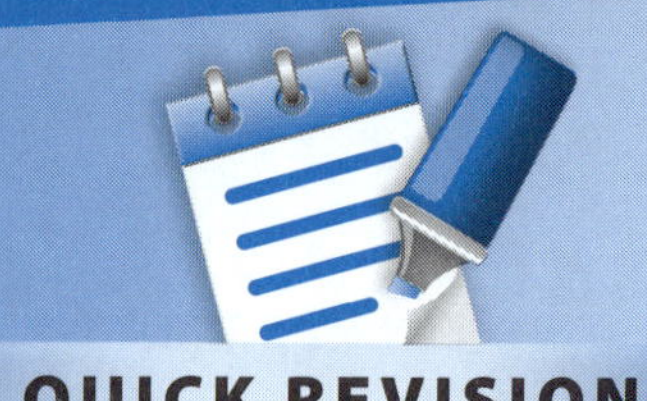

QUICK REVISION

1 Cells make up every part of __________ and plant bodies. Multicellular organisms have many different kinds of cells, each __________ for a different function. They require many __________ to grow and repair properly. Some of these nutrients are used by cells to provide __________, some become the building blocks for cellular products, and other nutrients are used in chemical reactions that are a part of cell __________ and repair.

The cell theory states that cells come from other __________. Some types of cells reproduce rapidly, while other cells don't reproduce at all. New cells are needed throughout life. These are for growth, to replace __________ and ageing cells and to repair __________ out tissues. As organisms grow, cells can only enlarge to a certain point. Then they __________, so a large organism consists of more cells than a smaller organism. Cells in large organisms are not necessarily any __________ than those in small organisms. The ability to reproduce allows them to repair parts. For example, damaged skin will simply reproduce allowing the organism to grow over and __________ the damage. Similarly, new bone grows on the surface of older bone to repair a __________.

In other cases it is a means of __________. For example, some single-celled organisms multiply by __________. For simple __________ (single-celled) organisms such as the amoeba, one cell division creates an entirely new organism as shown in the diagram below.

1 parent cell

2 nucleus divides

3 cytoplasm divides

4 two identical daughter cells are formed

2 In __________ (many-celled) organisms, cell division is used for growth and development, and to repair the organism. Mitosis is the process by which a cell separates the __________ in its nucleus into two identical sets, forming two separate __________. The primary result of mitosis is to transfer the parent cell's genes into two __________ cells. These two cells are genetically __________ and do not differ in any way from the original parent cell. Mitosis is an example of __________ reproduction.

3 Meiosis is a special type of cell division necessary for __________ reproduction in multicellular organisms. The cells produced by meiosis are called __________ (sex cells) or spores. In all animals and land plants, gametes are called __________ (male) and egg __________ cells. Meiosis does not occur in all cells of the body, but in specialised cells in the __________ organs.

Answers **1** animal; specialised (designed); nutrients; energy; growth; cells; damaged; worn; multiply; larger; repair; break; reproduction; dividing; unicellular **2** multicellular; chromosomes; nuclei; daughter; identical; asexual **3** sexual; gametes; sperm; female; reproductive

CELL DIVISION AND REPRODUCTION

Cells

REVISION SUMMARIES

1 New cells are needed throughout life. These are for growth, to replace damaged cells and to repair worn out tissues. For example, if your skin is cut, the cells along the edges of the cut undergo division to repair the cut. The same applies to broken bones. The cells along the edges of the break undergo division to repair the break.

Cell division is a way for cells to grow and reproduce. During cell division, an exact copy of the original cell is created. Even when an organism is fully grown, cell division continues to occur to renew damaged cells and in repair, replacing cells that die from normal wear and tear or from accidents. For example, dividing cells in bone marrow continuously make new blood cell. In humans, some 2.5 million new red blood cells are produced each second. These cells develop and circulate for about 100 to 120 days in the body before being broken down and their components recycled. This is because they become old or damaged as they squeeze through narrow blood vessels, or are bumped by other blood components or by the blood vessel walls as they travel. They are removed from the circulation by specialised cells (**macrophages**) found in the spleen and liver.

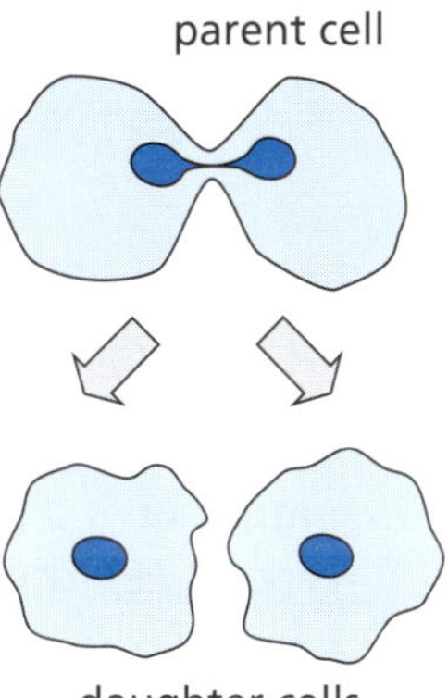

Cells are limited in size because the cell membrane must transport the food and oxygen to the parts inside. At the same time, it is through this membrane that wastes are eliminated. As a cell gets bigger, the increasing size of the cell membrane is unable to keep up with the volume inside the cell, because the inside grows at a faster rate. So, reaching a certain size, a cell replicates to form two smaller cells. The cells in the growing root of an onion are actively dividing. Observing some root cells under a microscope will show many of them in the process of **division**. Some cells, such as nerve cells, are unable to **replicate**.

2 **Mitosis** is the type of cell division used mainly for growth, repair and replacement of damaged cells and occurs in most of the cells of your body. Mitosis occurs wherever new cells are needed. It produces two cells that are identical to each other, and the parent cell from which they arose. Other cells, such as nerve cells and muscle cells, do not undergo cell division once they are fully formed.

Cell division is not exactly the same as mitosis. Mitosis is a precise division of the nuclear material in a cell, especially the **chromosomes**. This ensures that each daughter cell has the same inherited information (genes) and abilities as the parent cell from which it came. After mitosis has occurred, the cell itself is ready to divide. When this happens the two daughter cells are more or less identical. All the cells created through mitosis are **genetically identical** to one another and to the cell from which they came.

In mitosis, two daughter cells are produced from an original parent cell and these two daughter cells have the same number of chromosomes as the original parent cell. Chromosomes contain the coded or genetic material of a cell. There are 46 chromosomes in each normal human cell. The chromosome number of the daughter cells is the same as for the parent cell. A simplified example of mitosis involving two chromosomes is shown in the diagram on the next page.

The process of mitosis is quick and very complex. The whole sequence of events is broken up into stages corresponding to the end of one set of activities and the start of the next.

(cont.)

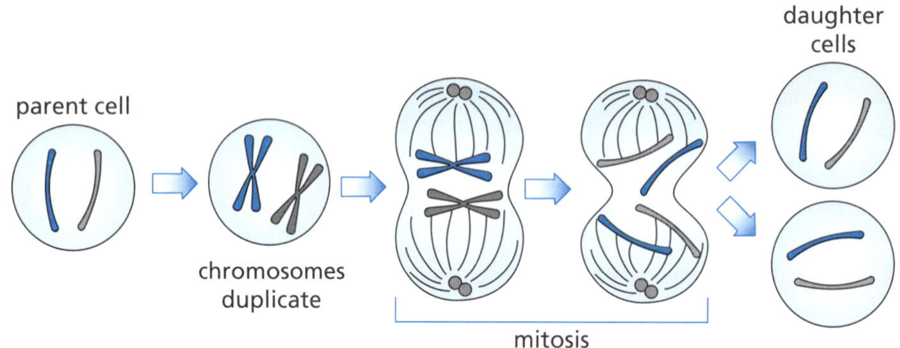

3 In sexual reproduction, two parents produce an offspring with a unique gene combination, each parent giving half of their chromosomes (and hence genes) to the offspring. **Meiosis** is a special type of cell division that produces **gametes** (sex cells) with half as many chromosomes. The male gamete is the sperm, the female gamete the ovum or egg. In human males about 200 million sperm are made each day, but in females generally only one viable ovum (egg) is made each month. When the two gametes fuse together during **fertilisation**, the resulting cell is called a zygote.

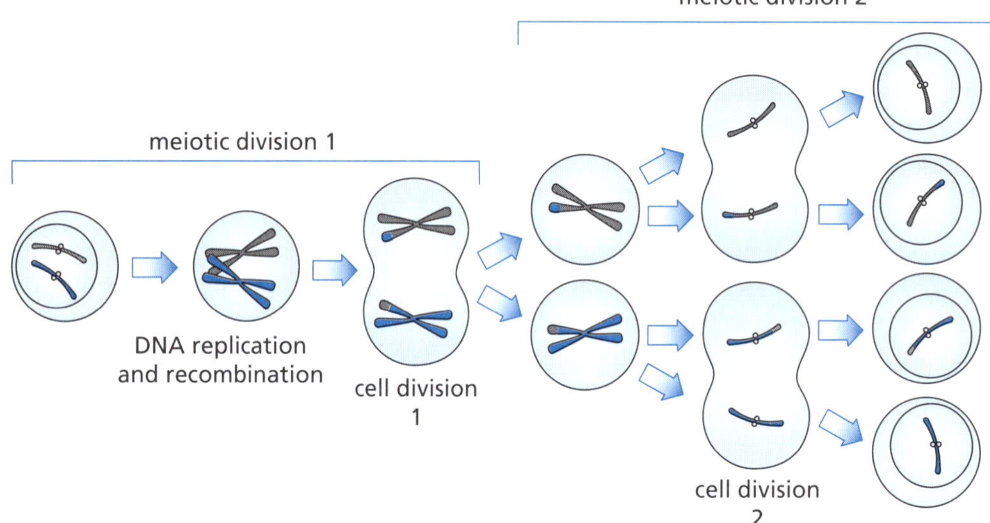

During meiosis, the chromosomes have the opportunity to exchange genetic material. This exchange of some genes is called **crossing over**. So the gametes formed are not identical to either themselves or to the parent cell. Cell division and gamete formation differs from kingdom to kingdom in multicellular organisms.

Checklist

Can you:

1 *Explain that new cells are needed throughout life for cell growth and repair?* ☐
2 *Describe the purpose of mitosis?* ☐
3 *Describe the function of meiosis and cellular reproduction?* ☐

CELL DIVISION AND REPRODUCTION

Cells

REVISION TEST

1 a Why must a cell divide? Why can't it just keep on growing and getting larger? *Hint 1* (3 marks)
  b What are the three functions of mitosis? (3 marks)
  c Is mitosis occurring in your body at the moment? (1 mark)

2 The following table is a summary of the daughter cells produced by mitosis and meiosis.

	Type of cells formed	Function of cells	Number of chromosomes	Genetic material
Mitosis	P	R	T	V
Meiosis	Q	S	U	W

Replace each of the letters in the table with one of the following statements. (8 marks)

First column: body (somatic) cells
sex cells (gametes)

Second column: sexual reproduction
growth, repair, asexual reproduction

Third column: half the number of the parent
same as for parent

Fourth column: same as for parent
half genetic material, with new gene combinations

3 A cell with 16 chromosomes undergoes mitosis.
  a How many daughter cells are created? (1 mark)
  b Each daughter cell has how many chromosomes? (1 mark)

4 When will cells generally divide by mitosis? (1 mark)

5 The following diagram shows one step during the process of mitosis.

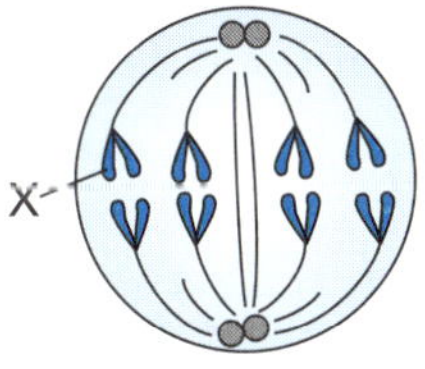

  a What structure is labelled X? (1 mark)
  b Why are these structures being dragged to opposite ends of the cell? (1 mark)

6 True or false? *Hint 2*
  a During meiosis the resulting gametes have the same number of chromosomes as the parent cell. (1 mark)
  b In meiosis, chromosomes can swap genes to produce unique gametes. (1 mark)
  c Meiosis is a type of cell division that produces sex cells. (1 mark)
  d Meiosis reduces the number of sets of chromosomes by half, so that when the gametes recombine during fertilisation the chromosome number of the parents will be re-established. (1 mark)
  e Mitosis and meiosis are the same processes. (1 mark)

(cont.)

f The daughter cells formed in meiosis are similar to the parent cells. (1 mark)

g There are four daughter cells formed in meiosis. (1 mark)

7 Mitosis is the process by which most cells in an animal's or plant's body are replicated. During replication a duplicate is made for each chromosome. The following diagram shows the steps during this process.

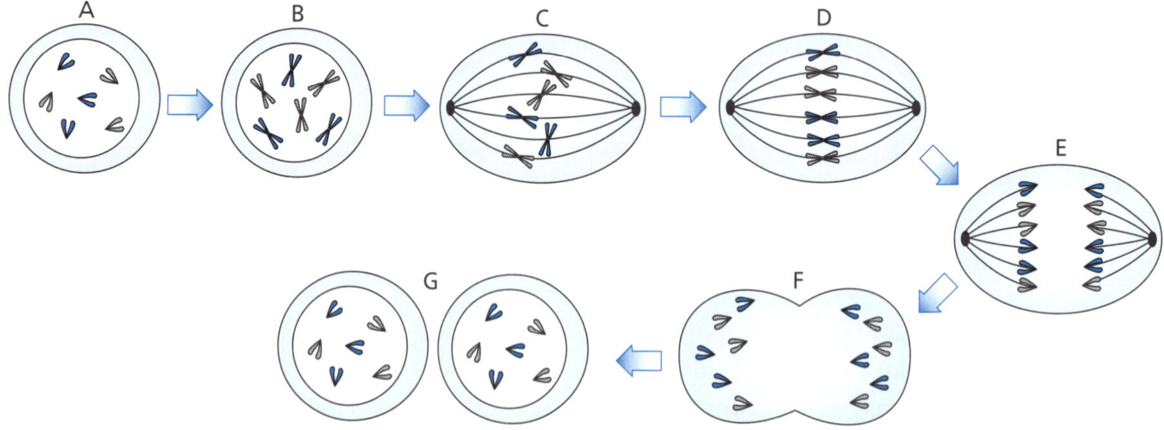

Replace each of the letters A to G with one of the following statements. *Hint 3* (7 marks)

- nuclear membrane disappears and microtubules attach to chromosomes
- the separated chromosomes are pulled apart
- chromosome number doubles
- microtubules disappear and the cell begins to divide
- two daughter cells produced, each with the same number of chromosomes as the parent cell
- resting phase where the cell has its full number of chromosomes
- chromosomes align in the middle of the cell

Hint 1: Compare the surface area of the cell with the volume of the cell as it gets larger.

Hint 2: These statements all refer to meiosis. Don't confuse it with mitosis.

Hint 3: Each of these statements is a description of what happens at each step.

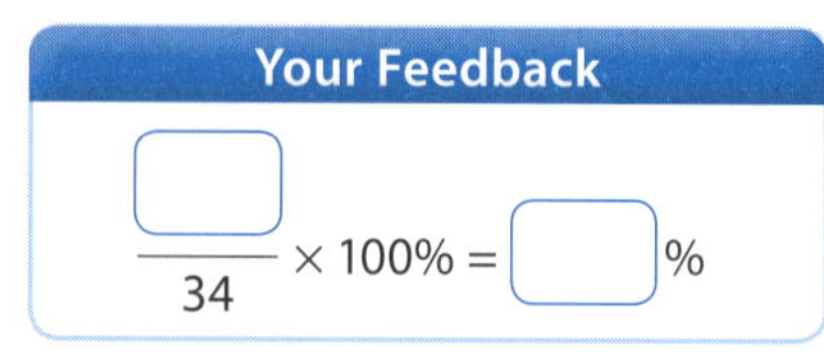

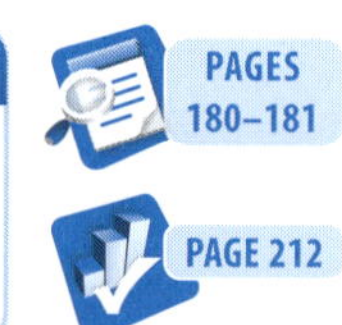

DISEASE

Cells

QUICK REVISION

1 Microbes are tiny __________, often too tiny to see without a __________. They are abundant on Earth and live everywhere; in air, soil, rock and water. Some __________ need oxygen to live, and some do not. These microscopic organisms are found in plants and animals as well as in the human body. Most microbes belong to one of five major groups: __________, viruses, fungi, algae or protozoa. Some microbes (called pathogens or __________ cause disease in humans, plants and animals. Others are __________ for a healthy life, and we could not exist without them. That microbes cause __________ diseases has been known since the 19th century. Towards the end of the 20th century researchers began to find out that microbes also contribute to many __________ (long-lasting) diseases and conditions.

2 There are several barriers __________ organisms from infection. These include __________, chemical and biological barriers. Our skin is an example of a mechanical barrier that is the __________ line of defence against infection. Similarly, the waxy cuticle of many leaves, the __________ of insects and the __________ and membranes of externally deposited eggs act as barriers to infection. But organisms are not completely sealed off from their __________ so other systems act to protect body openings such as the lungs, intestines and the urinary tract. In the lungs, coughing and __________ mechanically eject germs and other irritants from the respiratory tract. Tears and urine also expel germs, while __________ in the respiratory and gastrointestinal tract traps and entangles microorganisms.

Sometimes germs get past these defences and into our __________ and cells. The immune system is made up of a network of cells, tissues and organs that work together to __________ the body from infection and other invaders. It attacks organisms and __________ that invade body systems causing disease. Leucocytes are __________ blood cells that seek out and destroy disease-causing organisms or substances. The following image shows a leucocyte.

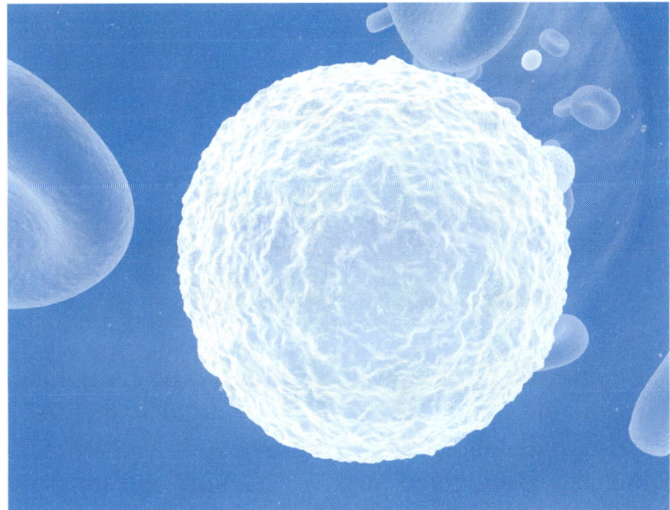

3 Antibiotics are powerful __________ that fight bacterial infections. They either kill bacteria or prevent them from __________. This supports your body's natural defences. When taking antibiotics, follow directions carefully and __________ the course of medicine even if you feel better. This is because each time you take antibiotics you increase the chances that bacteria in your body will develop __________ to them. Later, you could get or spread an infection that those antibiotics cannot cure. Antibiotics do not fight infections caused by __________.

(cont.)

__________, the first natural antibiotic, was discovered by Alexander Fleming in 1928. Vaccination or __________ is injecting a weakened or killed __________ in order to stimulate the body's immune system against the microbe, and so prevent __________. The healthy immune system is able to recognise invading bacteria and viruses and produce substances (__________) to destroy or disable them. Immunisation prepares the immune system to ward off a disease. The effectiveness of immunisations can be improved by periodic repeat injections or __________.

4 Non-infectious diseases are __________ due to disease-causing organisms (__________) and cannot be __________ from one person to another. They include genetic diseases (such as Down __________ and haemophilia), and those related to lifestyle or environment (such as cardiovascular disease, lung cancer and skin cancer). The following X-ray shows a patient with lung cancer.

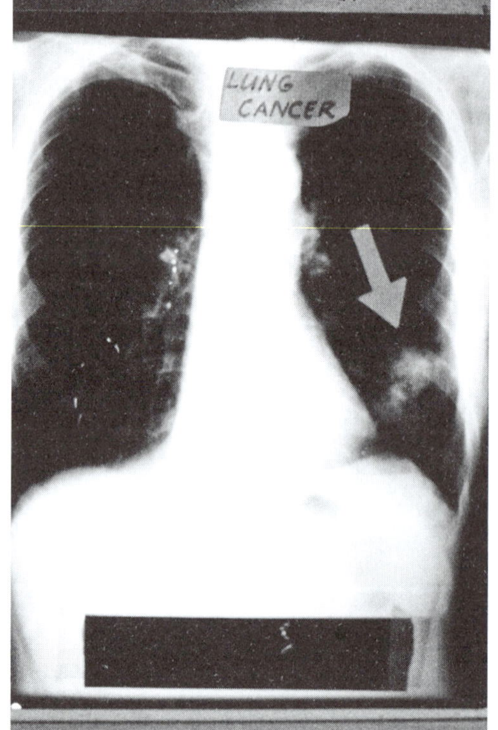

5 A healthy lifestyle is important. When we look after our physical health, we feel better, fitter, more relaxed and better able to __________ with things. Healthy living means maintaining a healthy lifestyle and introducing habits to improve your health. This includes regular __________, avoiding eating too much __________ food and sugary drinks and not smoking. Living a healthier lifestyle means there is a lower risk of developing many __________.

Answers 1 organisms; microscope; microbes; bacteria; germs; essential (necessary); infectious (contagious); chronic 2 protecting; mechanical; first; exoskeleton; shells; environments; sneezing; mucus; bloodstreams; protect; substances (chemicals); white 3 medicines; reproducing; finish (complete); resistance; viruses; penicillin; immunisation; microbe; disease; antibodies; boosters 4 not; pathogens; shared (spread); syndrome 5 cope; exercise; junk; illnesses (diseases)

1 Microorganisms (microbes) are very small and generally cannot be seen without a microscope. They include **viruses**, **bacteria**, **algae**, **protozoa** and **fungi**. Most microbes are useful. They break down dead animal and plant matter into simpler substances so it can enter the beginning of a food chain. They also break down sewage and other wastes, fix gases so plants can use them and are exploited to produce foods and medicines.

Only a small number of microbes cause disease, and these are called **pathogens**. They can invade and multiply inside an individual causing an **infection**. For example, the disease called meningitis can be caused by various microbes. Meningitis is an inflammation of the tissues covering the brain and spinal cord. This inflammation can be caused by viruses and fungi, as well as bacteria. While viral meningitis is the most common type, it has no specific treatment but is usually not as serious as meningitis caused by bacteria.

The following diagram shows the cycle of transmission of disease.

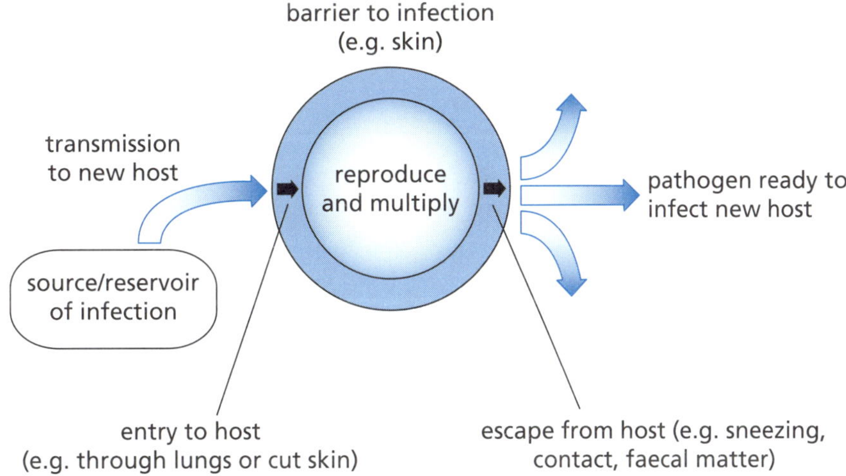

Microbes can be transmitted:

- directly, such as through kissing, touching, sneezing or through breastmilk
- indirectly, such as through air, food, water or insect bites.

The germ theory of disease (also known as the pathogenic theory) proposes that microorganisms cause many diseases. It was highly controversial when first proposed, but the germ theory was proved to be true in the late 19th century. It is now a fundamental part of modern medicine and led to **antibiotics** and hygienic practices.

2 Organisms have **barriers** to prevent microbes from entering. There are physical and chemical barriers on the surface of the body and specialised cells and molecules inside it. The skin is the most important barrier to infection. There are many microbes living on our skin but these rarely penetrate it. Damage to the skin, such as a cut, allows pathogens to penetrate the underlying tissues. The respiratory tract is also exposed to pathogens present in the air. This tract is lined with sticky mucus secreted by cells that traps pathogens.

Sometimes pathogens get past these barriers. The **immune system** must detect a wide variety of these agents and distinguish them from the organism's own healthy tissue. White blood cells, or leucocytes, are cells of the immune system involved in defending the body against both

(cont.)

infectious disease and foreign materials. The two basic types of leucocytes are:

- phagocytes, which are cells that eat (engulf) and destroy invading organisms
- lymphocytes, which are cells that allow the body to remember and recognise previous invaders and help the body destroy them.

There are a number of different lymphocytes that have different jobs in fighting infection. For example, when **antigens** (foreign substances that invade the body) are detected, they trigger certain lymphocytes to produce antibodies. These are specialised proteins that lock onto specific antigens. The **antibodies** continue to exist in an organism's body, so that if the same antigen is presented to the immune system again it is ready to fight it and the person doesn't get sick from it again. This is why those people who have had an active measles infection or who have been vaccinated against measles have immunity to the disease. There are also other lymphocytes that destroy antigens which have been tagged by these antibodies. Antibodies also can also neutralise toxins (poisonous or damaging chemicals) produced by different organisms.

The following table shows some diseases and infections caused by microbes. You do not need to remember this table; you just need to be aware that some common diseases may have several causes.

	Bacteria	Fungus	Protozoa	Virus
athlete's foot		✓		
chickenpox				✓
common cold				✓
diarrhoea	✓		✓	✓
influenza				✓
malaria			✓	
pneumonia	✓	✓		✓
tuberculosis	✓			
urinary tract infection	✓			

3 **Active immunity** can also be generated artificially, through **vaccination** and **immunisation**. This can prevent certain diseases. An antigen is introduced into the body in a way that doesn't make someone sick, but allows the body to produce antibodies. This will then protect the person from future attack by the germ or substance that produces that particular disease.

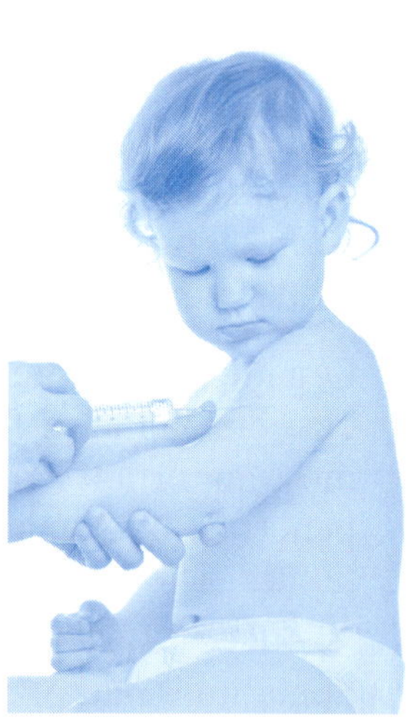

Antibiotics are drugs used to treat infections caused by bacteria. These include such illnesses as tuberculosis, salmonella, syphilis and some forms of meningitis. They are not effective against viruses. An antibiotic works in one of two ways.

- It can kill the bacteria. This antibiotic interferes with how the bacterium forms its cell wall or its cell contents. Penicillin is an example.
- It stops bacteria from multiplying.

With the development of antibiotics and vaccination programs, infectious disease is no longer the leading cause of death in the western world. The photograph shows a child being vaccinated.

4 Some diseases are **non-infectious** (also called non-communicable). They are not caused by a pathogen and cannot be transmitted from one person to another. Some may be chronic (long-lasting or recurring) diseases, or they may result in more rapid death. Examples of non-infectious diseases include autoimmune diseases, heart disease, stroke, many cancers, asthma, diabetes, chronic kidney disease, osteoporosis, Alzheimer's disease and cataracts. The following diagram compares healthy bone and a bone with osteoporosis.

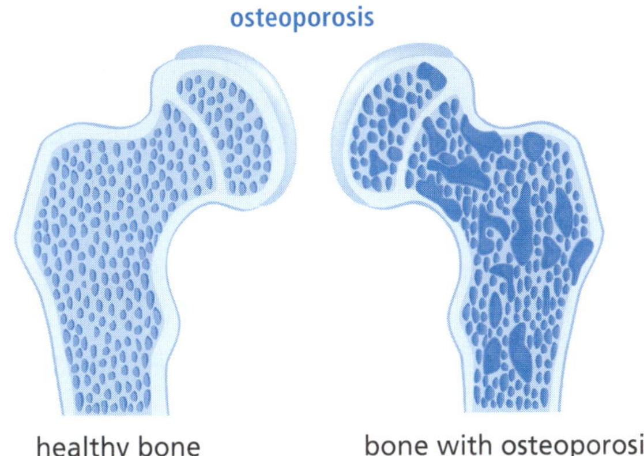

5 Maintaining and promoting health is achieved through different combinations of physical, mental and social wellbeing (sometimes referred to as the health triangle). People can improve their health by exercising, getting enough sleep, eating wholesome and nutritional foods, maintaining a healthy body weight, not abusing alcohol and avoiding smoking and other non-**therapeutic** drugs. The **environment** is an important factor influencing health. Factors such as clean water and air, adequate housing and safe communities contribute to good health, especially of infants and children. If a person is run down, stressed or malnourished, infectious diseases have a greater chance of taking hold.

Checklist

Can you: ✓

1. *Define infectious disease and name the organisms that cause it?* ☐
2. *Identify barriers to infectious disease?* ☐
3. *Describe the purpose of antibiotics and vaccination?* ☐
4. *Identify and describe non-infectious diseases?* ☐
5. *Identify the factors for good health?* ☐

DISEASE

Cells

REVISION TEST

1 Around the outbreak of the Peloponnesian War in ancient Greece, an epidemic occurred in Athens from 430 to 426 BC, killing many people and contributing significantly to the decline and fall of classical Greece. Descriptions of this infectious disease indicate it could have been typhus or smallpox.

a Suggest a reason why this epidemic spread so quickly. *Hint 1* (1 mark)

b Why is such an epidemic of these diseases unlikely to occur today? (2 marks)

2 Tuberculosis (TB) has been present in humans since ancient times. *Mycobacterium tuberculosis* causes the disease tuberculosis. It is spread from person to person through the air. A person with this disease only needs to cough or sneeze and fine aerosol droplets loaded with this organism are spread. Eventually the moisture evaporates, but the very light organisms can remain airborne for days and spread over large distances. *Mycobacterium* has a waxy coat and this protects it from drying out.

a What type of organism causes tuberculosis? (1 mark)

b Explain how the mode of transport of tuberculosis can be direct and indirect. (2 marks)

c How is *Mycobacterium* able to survive for many months in dust and air, and can withstand weak disinfectants? *Hint 2* (1 mark)

d Most deaths from tuberculosis occur in developing countries. Suggest one reason. (1 mark)

e 'Tuberculosis is closely linked to both overcrowding and malnutrition, making it one of the principal diseases of poverty.' How can overcrowding and malnutrition increase the chance of getting infected? (2 marks)

f What type of drug is used to treat tuberculosis? (1 mark)

3 The skin is one line of defence preventing the entry of pathogens. Explain how the following can prevent entry of disease-causing organisms.

a urine (1 mark)

b pathogens in food (1 mark)

c earwax (cerumen) (1 mark)

4 a How does the body recognise when an infection has taken place? (1 mark)

b There are various types of white blood cells. Give two examples of their functions. (2 marks)

5 Most upper respiratory tract infections, such as the common cold and sore throats, are generally caused by viruses. A doctor prescribes a course of antibiotics. Will this work? (1 mark)

6 Explain the differences between infectious and non-infectious diseases. (4 marks)

7 A non-infectious disease is a disease that may be caused by the environment, nutritional deficiencies or genetic inheritances. Which of these categories would each of these diseases belong to?

a skin cancer from ultraviolet radiation (1 mark)

b anaemia from not enough iron, folate and vitamin B12 in the diet (1 mark)

c scurvy from lack of vitamin C (1 mark)

d sickle-cell anaemia caused by a mutation in the haemoglobin chain (1 mark)

e cirrhosis, chronic liver damage from long-term alcohol abuse (1 mark)

f lung cancer due to smoking (1 mark)

g haemophilia due to a defective gene (1 mark)

8 True or false?

a Colds and flu are essentially the same thing. (1 mark)
b Antibiotics are used to treat colds and flu. (1 mark)
c Antibiotics should only be taken for as long as you feel symptoms. (1 mark)
d Meningococcal disease is contagious. *Hint 3* (1 mark)
e Cold weather carries viruses to infect people with colds more easily. (1 mark)
f Handwashing is one of the best possible ways to prevent the spread of germs. (1 mark)
g Infectious diseases are caused by tiny organisms called parasites. (1 mark)
h Mesothelioma, a cancer caused by exposure to asbestos, is a non-infectious environmental disease. (1 mark)

9 The following graph shows the incidence of measles in a certain country over a number of years.

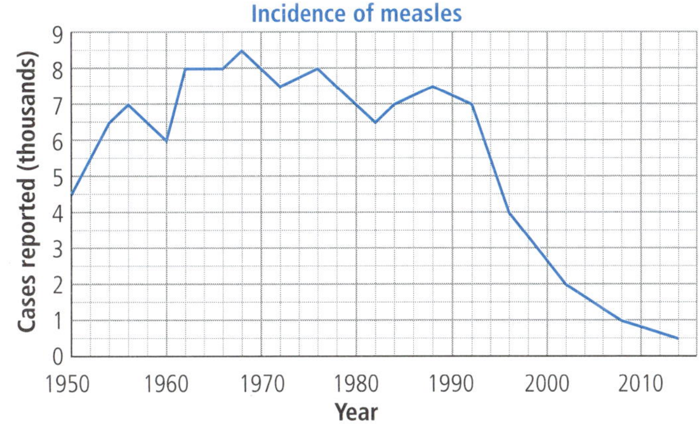

a In which year shown did the number of reported cases first reach 7000? (1 mark)
b When were the most cases reported? How many were reported in that year? (2 marks)
c In which year did the number of reported cases drop below 4000 cases? (1 mark)
d The government, concerned over the high incidence of measles, instituted a wide use of the measles vaccine. In which year did this most likely begin? (1 mark)
e Predict how many cases might be reported by 2020. (1 mark)
f What is the result of the widespread use of the measles vaccine? (1 mark)

Hint 1: Consider the conditions and knowledge that people had at the time.
Hint 2: What features might the bacterial cell wall have to give it an advantage?
Hint 3: Meningococcal disease is caused by the bacterium Neisseria meningitidis. Does this help?

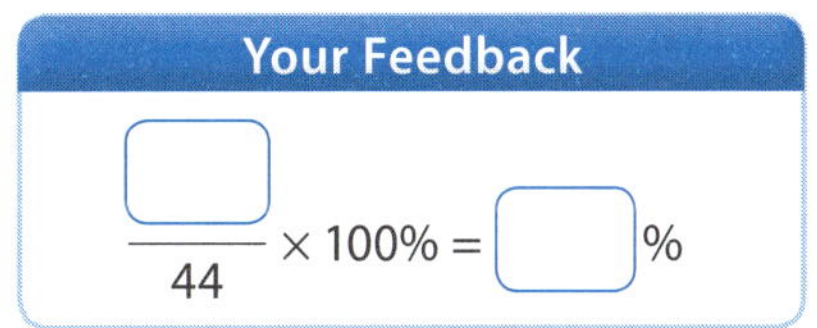

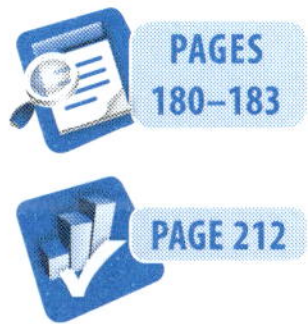

MULTICELLULAR ORGANISMS

Body systems

QUICK REVISION

1 Animals and plants are ________ as multicellular organisms. In these organisms, various cells take on specific roles. Some cells specialise in the absorption of oxygen which is used for ________ respiration. Waste elimination, nutrient absorption and reproduction are carried out by other cells. In order to perform these tasks, similar cells are organised into ________. In turn, different tissues are organised to form a specific ________. The whole organism consists of a system of organs working together. In humans the ________ system consists of the kidneys, bladder, ureters and the urethra. The circulatory system consists of the heart, arteries, veins and capillaries through which blood flows. In flowering plants, the organ system consists of roots, ________, leaves and flowers. In each of these organs, various tissue types are present. Multicellular organisms, therefore, are more ________ than single celled protists.

2 Humans are vertebrates. In all vertebrates there are many ________ types of tissues. Muscle tissue, skin tissue and nervous tissue are some examples. Organs are composed of a number of different ________. Muscles, for example, contain not only muscle tissue but also nerves and blood vessels. The digestive system consists of many organs including the stomach, liver, small intestine and large intestine. Microscopic examination of the tissues of each organ shows that the cells have different ________ and sizes.

The largest organ in humans is the ________, which has a mass of about 3 kg. The epidermal layers of the skin ensure that your body is waterproof and also act as a barrier to attack by ________. The skin must be supplied with ________ and nutrients to keep it alive. These arrive via the blood capillaries and wastes are removed in turn via these capillaries. The skin has many ________ that allow us to feel pressure, heat and pain. On hot days the skin helps to regulate temperature control by releasing sweat from sweat glands. As the sweat evaporates, the body is ________.

3 Plants are also multicellular organisms. There is a wide variety of plants including flowering plants, conifers and ferns. All these plants have organs called roots, stems and ________. In terms of reproduction, the flowering plants have flowers as organs of reproduction. Conifers reproduce using ________ and ferns via spores. As in animals, each organ is ________ of a variety of different tissues. In the root of a plant there are tissues called ________ that are involved in water absorption from the soil and the ________ of that water up the plant stem. The root requires nutrients and oxygen to stay alive and these are transported in ________ tissues from the leaves via the stem. Other cells in the cortex store nutrients including carbohydrates such as starch. The epidermis is the outer layer of cells of the root. Some of these cells have extensions called root ________ which absorb the water from the soil.

Answers **1** classified; cellular; tissues; organ; excretory; stems; complex **2** different; tissues; shapes; skin; microbes; oxygen; nerves; cooled **3** leaves; cones; composed; xylem; transport; phloem; hairs

MULTICELLULAR ORGANISMS

Body systems

REVISION SUMMARIES

1 As a living organism, you are able to grow and move, take in food and oxygen, eliminate wastes, detect (through your senses) the environment and respond to it, and are able to reproduce. If you were a single-celled organism, all these different functions would be done by different parts of your one cell. Your body, however, is much more complex as it is multicellular (consists of many cells). In a **multicellular** organism, cells specialise. These different types of cells form tissue. Groups of **tissues** form **organs** and different organs form **organ systems**. For example, in the excretory system there are different organs such as the kidneys and bladder.

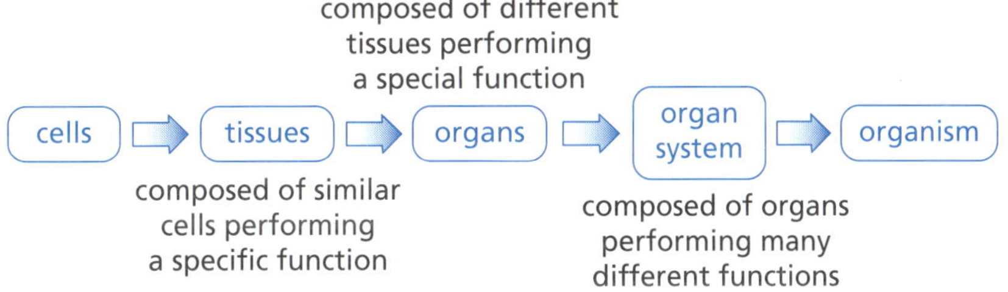

2 **Vertebrates** are multicellular animals. Many different cells form a wide variety of tissues. These in turn form organs. The largest organ is the skin. The skin has a total surface area of about 1.7 m^2 and weighs 3 kg. The skin is made up of many different tissues, as shown in the following diagram. The epidermis is the outer layer of the skin. The dermal layer consists of sublayers of tissues. **Blood capillaries** can be found in this dermal layer as well as fat cells, sweat glands and hair follicles. The skin guards the internal organs of the body as well as the bones and muscles. The skin acts as a barrier to the attack of **pathogenic organisms**. It is waterproof and helps to regulate the heat and moisture balance of the body. Exposure to UV radiation can lead to sunburn, skin ageing and the formation of skin cancers.

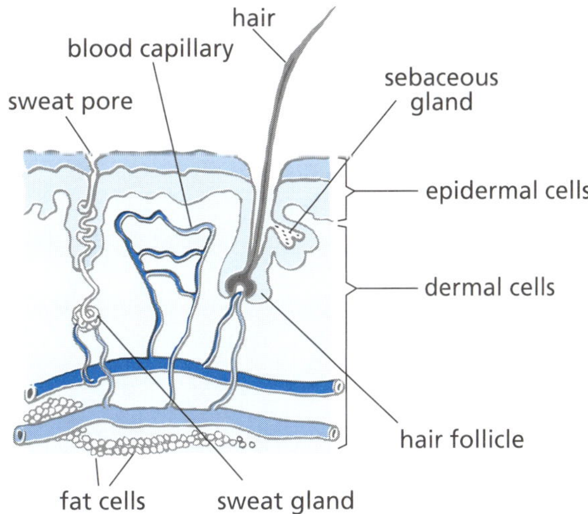

3 Green plants also contain different cells, tissues and organs. The major organs of flowering plants, conifers and ferns are **roots**, **stems** and **leaves**. In flowering plants (**angiosperms**) the flowers are the reproductive organs. All these organs contain a variety of different tissues. In the plant stem we find structures called vascular bundles that contain xylem, phloem and cambium cells.

(cont.)

- **Xylem cells** form long tubes that transport water from the roots, through the stem and into the leaves.
- **Phloem cells** form long tubes that transport nutrient solutions (e.g. containing glucose) from the leaves down the stem and into the roots.
- **Cambium** is a thin layer of cells between the xylem and phloem tissue. These cells actively divide to create new xylem and phloem cells.
- **Cortical** and some other cells are unspecialised cells that support the stem and store some nutrients such as starch.

vascular bundles in a plant system

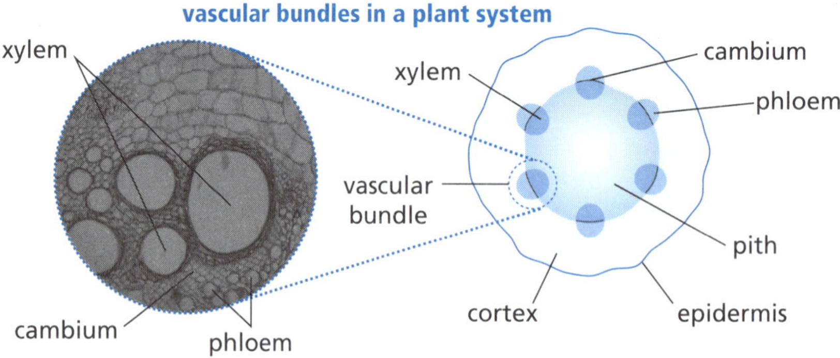

Checklist

Can you:

1. *State the organisational levels of cells in a multicellular organism?* ☐
2. *Describe the tissues that make up skin?* ☐
3. *Describe the function of the organs and tissues of a green plant?* ☐

MULTICELLULAR ORGANISMS

Body systems

REVISION TEST

1 The trunk of a woody tree, such as the eucalyptus, is an important organ. It contains a number of different cells and tissues. The following diagram shows a transverse sector of such a tree.

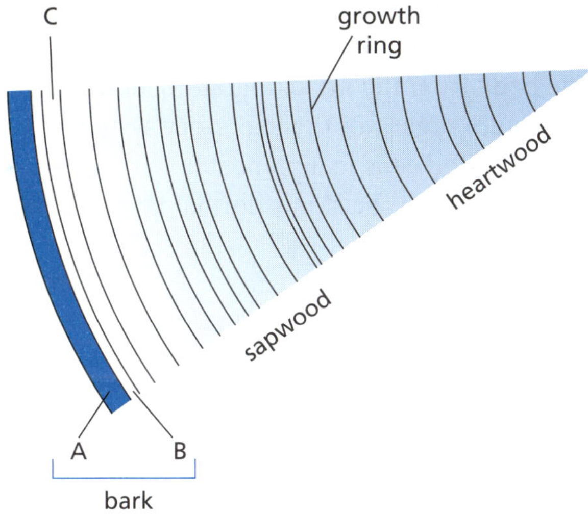

- **a** The bark is composed of two types of tissues (A and B). One tissue transports nutrients down from the leaves whereas the other tissue is a protective layer for the trunk. This layer is called cork.
  - **i** Which tissue is cork? (1 mark)
  - **ii** Name the other tissue. (1 mark)
- **b** Which tissue contains actively growing cells? Name this tissue. (2 marks)
- **c** The number of growth rings tells us the age of the tree. What is the age of the tree in the diagram? (1 mark)
- **d** Farmers used to clear trees from their land by ringbarking them. A deep circle of tissue was cut out around the tree trunk. This procedure killed the tree by stopping nutrients from reaching the roots. Which layers of the tree would need to be cut into to ringbark it? (1 mark)
- **e** The sapwood and heartwood contain another vascular tissue. Name this tissue and state its function. *Hint 1* (2 marks)

2
- **a** Multicellular organisms are composed of many different types of cells arranged in groups called tissues.
  Explain this statement. (1 mark)
- **b** Give an example of such a cell and the tissue to which it belongs. (2 marks)

3 A stem of fresh celery is placed in a jar that is half-full of red-dyed water. After several hours the stem is removed and transverse sections of the stem are removed and examined under a binocular microscope. The diagram on the right shows the result of the experiment.

State the aim of this experiment and state one conclusion that could be made. *Hint 2* (2 marks)

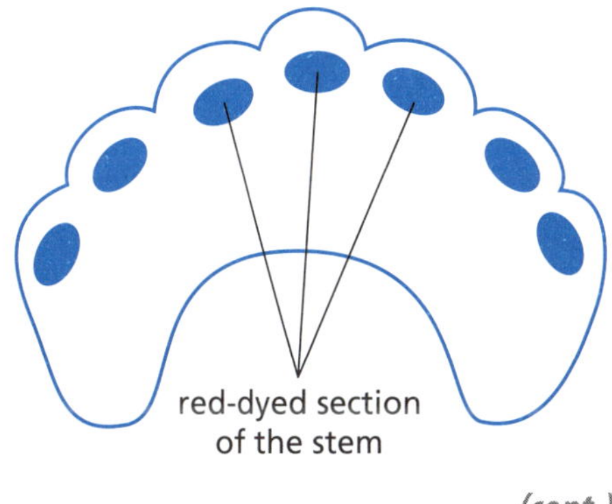

(cont.)

4 True or false?

a The heart is an organ system. (1 mark)

b An organ is composed of different tissues. (1 mark)

c The skin is a tissue composed of many cell types. (1 mark)

d The skin protects the body from the attack of pathogenic microbes. (1 mark)

e The kidney and bladder are organs of the excretory system. (1 mark)

f Cortical cells are unspecialised cells in a plant stem. (1 mark)

g Blood capillaries can be found in the skin's dermal layer. (1 mark)

h Fat cells are present in the skin. (1 mark)

i Vertebrates have organs that eliminate waste. (1 mark)

j Sweat pores are located in the dermal layer of the skin. (1 mark)

5 Our skin can be readily damaged by extended exposure to UV radiation from the Sun. The following table shows how skin burn can be reduced by the application of sun protection factor (SPF) sunscreens.

SPF	0	2	3	4	5	10	20	40
% reduction in sunburn	0	50	68	75	80	90	95	98

a Plot this data as a line graph of best fit. *Hint 3* (5 marks)

b Estimate the percentage reduction in sunburn if the SPF is 15. (1 mark)

c Identify two other options for avoiding sunburn. (2 marks)

Hint 1: Vascular tissue refers to fluid transport tubes.
Hint 2: The dyed water enters the vascular bundles.
Hint 3: Make sure the data occupies a minimum of 80% of the grid space. Your line of best fit should be a smooth curve.

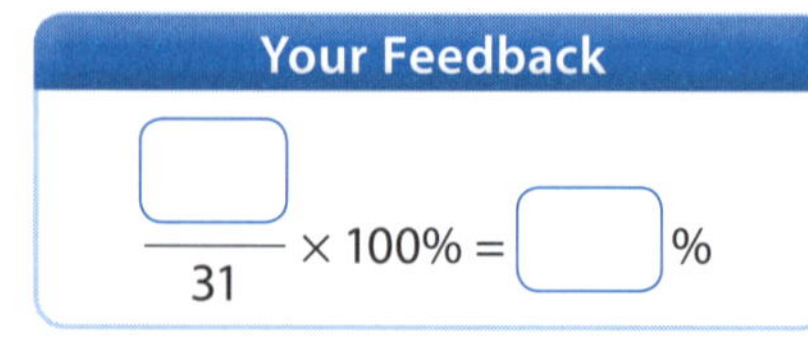

DIGESTIVE SYSTEM

Body systems

QUICK REVISION

1 Cells require nutrients and ________ in order to live. The nutrients come from the food we eat and the oxygen comes from the ________ we breathe. It is the role of the digestive system to break down food into usable nutrients. It is the role of the respiratory and circulatory systems to bring the oxygen to each cell. Your food contains molecules of varying sizes and these must be broken down to ________ sizes so they can be used by the body. This is one of the roles of the digestive system. Some food components cannot be used and this forms the waste that is removed from the body as ________. Two important organs are the liver and pancreas. The liver has many roles. It can store excess food ________ as polymeric molecules called glycogen. The liver also makes ________ which is used to convert fat into tiny droplets so it can be digested more readily. The pancreas, which is located just below the ________, produces digestive enzymes which enter the small intestine as partly ________ food enters the intestine after ________ the stomach. The pancreas is also involved in the metabolism of sugars.

2 The teeth are an important component of the digestive system. Teeth are usually classified into ________ categories based on their shape and function. The incisor teeth are at the front of the mouth and they are used to ________ the food. The next teeth are the sharp ________ teeth which rip and tear the food. Next to them are the pre-molars and molars. Their job is to ________ and grind the food. Molars are flat-topped with ridges on their surfaces. Because humans are omnivores, we eat food derived from both ________ and animals and so our teeth are less ________ than those of carnivores or herbivores. ________ use their canines to grab, rip and tear their food and so they have sharp canines and sharp molars. A herbivore needs to ________ its food for long periods of time and so needs large premolars and molars. They usually do not have any canine teeth.

3 The alimentary canal is a long ________ which runs from the mouth to the anus. Food is placed in the mouth where it is ________ down by your teeth and mixed with saliva from the salivary glands. The first step of digestion occurs in the mouth as salivary ________ attack the carbohydrates in the food. The food mixed with saliva then moves down your ________ into the stomach. Muscular contractions of the oesophagus wall help to ________ the food bolus down. The food that enters the stomach is attacked by gastric juices that contain ________ and enzymes. The proteins undergo significant digestion whereas the fats and carbohydrates do not. The soupy mixture that forms in the stomach then passes into the top of the small intestine which is called the ________. In the duodenum the partly digested food is mixed with bile (which has been stored in the ________ bladder) as well as digestive juices from the pancreas. The digestive process is completed in the small intestine and small nutrient molecules ________ into the blood stream through the intestinal wall. The material that remains now moves into the large intestine where ________, minerals and vitamins are absorbed into the blood. The semisolid waste is stored in the rectum at the end of the large intestine before it is passed out through the anus. Carnivores have similar digestive systems to humans. Ruminant animals such as sheep and cows have a more complex stomach that is ________ into four compartments. The cellulose fibres of the grass they eat take a long time to digest. Microbes ________ the food in the first two compartments before regurgitation back into the ________ where further chewing occurs. It is swallowed again to allow further digestion to occur.

Answers **1** oxygen; air; smaller; faeces; protein; bile; stomach; digested; leaving **2** four; cut; canine; crush; plants; specialised; carnivores; chew **3** tube; broken; enzymes; oesophagus; push; acid; duodenum; gall; diffuse; water; divided; ferment; mouth

1 Every cell in your body needs **food** and **oxygen** to survive. The food you eat cannot be used by the cells until it has gone through the stages of the digestive process. Only when your food is reduced to simple chemicals, small enough to pass through the tiny channels in the intestine walls, is it able to be absorbed into the blood and transported to all cells for growth, energy and repair. Any unused food, plant fibres and lots of water pass into the large intestine where the water is reabsorbed, leaving semisolid material (faeces) which is excreted.

The digestive process takes place in your digestive system: the **alimentary canal**, the **liver** and the **pancreas**. The **alimentary canal** is a tube running from mouth to anus. The liver, a large red-brown organ, is a chemical factory with many jobs. The liver changes digested proteins not needed by the body into glycogen and stores it. It makes a green fluid called bile that helps break up fats. It removes iron from old blood cells and stores it so it can be used later. It also removes some poisons and drugs from the blood system. The pancreas, a pale coloured organ, produces digestive enzymes that help break down food into smaller pieces. Other enzymes are made in different parts of the digestive system.

2 Vertebrates have two tasks in the process of taking in food. The first is finding a suitable supply of it and the second is being able to eat it. Once the food is in the mouth, **teeth** are used to begin the long process of digestion. Most mammals possess teeth, but these vary greatly from species to species. Mammals generally start to break down food by gnawing, grinding or tearing it, and their teeth have become adapted to these different requirements. Rodents (e.g. rabbits, rats, squirrels and guinea pigs) gnaw their seeds, nuts or roots. Their front teeth, the **incisors**, are sharp and prominent, continuing to grow throughout the life of the animals. Elephants and hippopotamuses also have long incisors, but they are used as tools, not for gnawing. Elephants' long incisors are their tusks. They are used for attack and defence, and for digging roots from the ground and breaking limbs from trees. Hippopotamuses use their incisors like a fork to direct plant material into their mouths.

Herbivores (e.g. cows, horses, sheep and kangaroos) are plant eaters. They grind their food. Their **premolars** and **molars** are flat-topped and have grinding surfaces to break up the tough grass and plant cell structures. **Carnivores** (e.g. cats, dogs, lions and wolves) are meat eaters. They have reduced incisors and large, prominent sharp **canine** teeth, as well as pointed premolars and molars. Carnivores tear off chunks of flesh and swallow without chewing. The last group of animals are the **omnivores** (e.g. pigs, humans, bears and foxes). Omnivores eat all types of foods. They have teeth that are not highly adapted. The following diagram shows the teeth of various vertebrates.

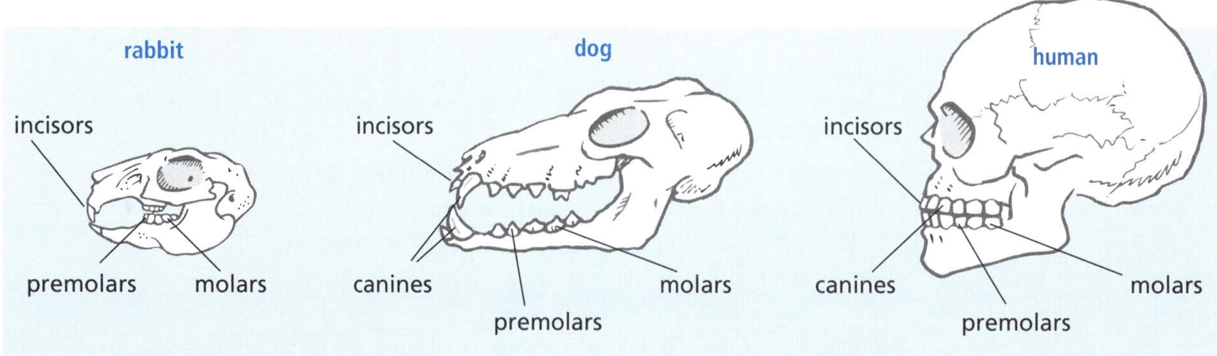

3 The diagram shows the organs of the human **digestive system**. When food is swallowed, it passes into the food tube or **oesophagus**. The oesophagus contains muscles that squeeze the balls of food into our stomach.

Unlike humans, the stomach of a cow or a sheep is divided into four compartments. The **cellulose** in the grass is difficult to digest and so it needs to spend a lot of time in the stomach. Initially, the chewed grass is fermented by microbes in the first two compartments of the stomach. The partly digested grass (cud) is often sent back up to the mouth for re-chewing. This is why you often see cows standing in the field just chewing and not tearing off more grass from the ground. The cud is again swallowed and further digested by microorganisms before passing into the intestine.

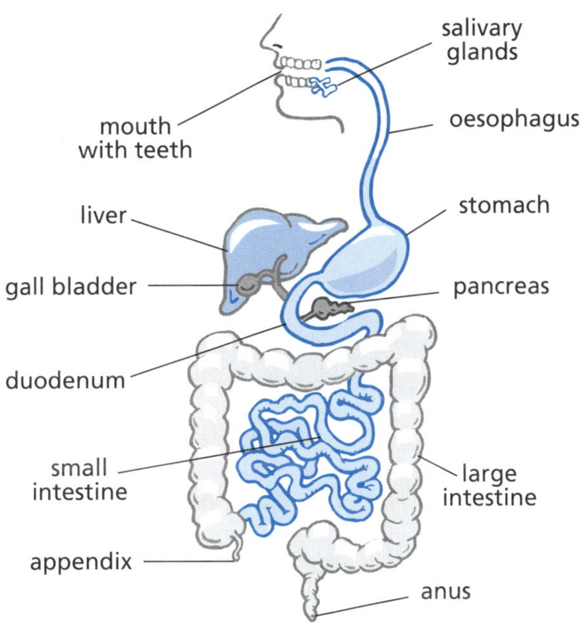

The following table describes the functions of the major parts of the alimentary canal.

Mouth	Stomach	Duodenum	Small intestine	Large intestine
• Teeth grind up food into smaller pieces. • Enzymes from the salivary glands start to attack and digest carbohydrates. • Proteins and fats are not digested.	• The food that is swallowed passes down a tube called the oesophagus and into the stomach. • The muscles of the stomach wall churn food with acids and enzymes into a soupy mixture. • Carbohydrates and fats are not further digested but proteins start their digestion into smaller molecules.	• This is the first part of the small intestine. • Bile from the liver (which has been stored in the gall bladder) emulsifies fats to form small fat droplets. • Enzymes from the pancreas attack fats, proteins and carbohydrates. • Carbohydrates are broken down into smaller molecules like sugars.	• Digestion is completed. • Digested nutrients diffuse into the bloodstream through the intestinal wall. • The intestinal wall has many small projections called villi which increase the surface area and increase the rate of diffusion.	• Water and any vitamins and minerals that remain are absorbed into blood. • Waste is transferred to the rectum before being excreted.

Checklist

Can you:

1 *Describe the purpose of the digestive system?* ☐
2 *Name the major organs of the digestive system and explain their function?* ☐
3 *Compare the role of different teeth in the feeding process?* ☐

DIGESTIVE SYSTEM

Body systems

REVISION TEST

1 The following diagram shows the skull of an animal.

a Identify the teeth labelled X and Y. (2 marks)

b Is this animal a carnivore, herbivore or omnivore? (1 mark)

c Identify the animal from this following list: cat; dog; sheep; pig. *Hint 1* (1 mark)

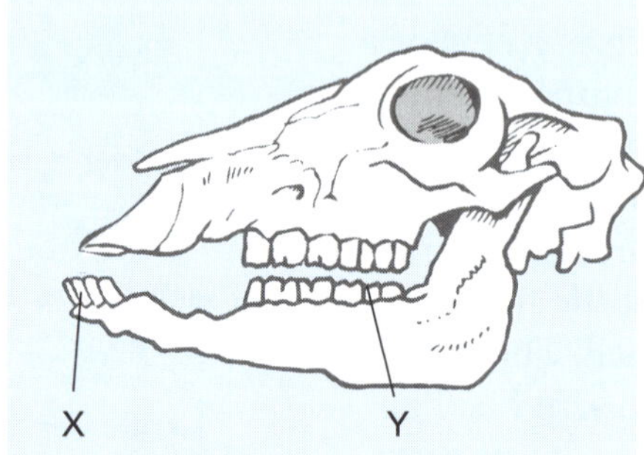

2 The following diagram shows the intestine and a blood vessel. Use the diagram to write a statement describing how food passes into the bloodstream. (2 marks)

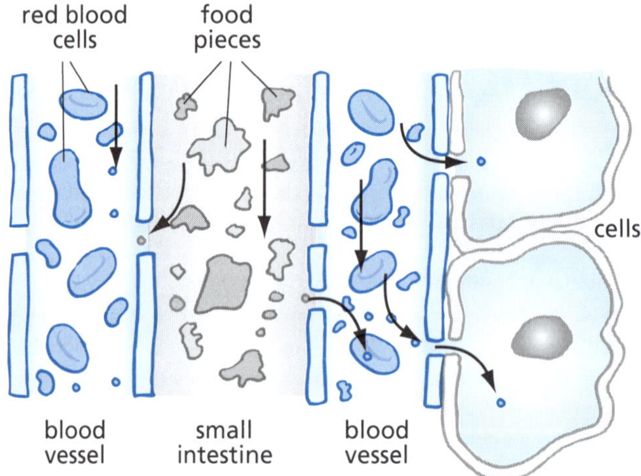

3 The statements below all describe what happens at different parts of the digestive system. Match each of the following words with one of the statements.

mouth stomach duodenum small intestine large intestine

a Water, vitamins and minerals absorbed. *Hint 2* (1 mark)

b Bile from the liver and gall bladder emulsifies fat into small droplets. (1 mark)

c Food is mixed into a soupy mixture with acids and enzymes. (1 mark)

d Food is mechanically broken down into smaller pieces. (1 mark)

e Digested nutrients diffuse through the wall into the bloodstream. (1 mark)

DIGESTIVE SYSTEM

4 Use the code letters in the following diagram to label the organs of the human digestive system. (8 marks)

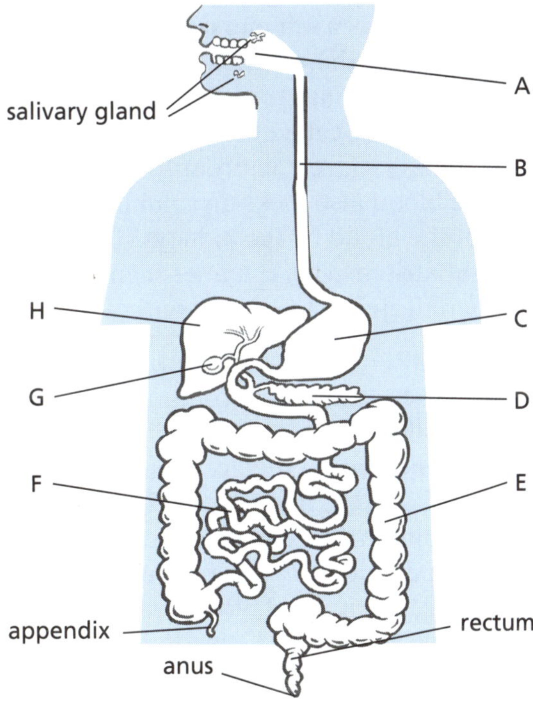

5 Before food enters the stomach, it has (to some extent) been crushed and reduced in size.

a Where does this occur? What processes occurs there? (4 marks)

b Explain the following basic functions of the stomach using only one sentence for each.

i short-term storage reservoir *Hint 3* (1 mark)

ii substantial chemical and enzymatic digestion initiated (1 mark)

iii vigorous contractions of stomach wall muscles (1 mark)

iv food slowly released into the small intestine (1 mark)

6 The following diagram shows a simple water-dwelling invertebrate called a hydra. Name two features of its digestive system which are different to that of a human. (2 marks)

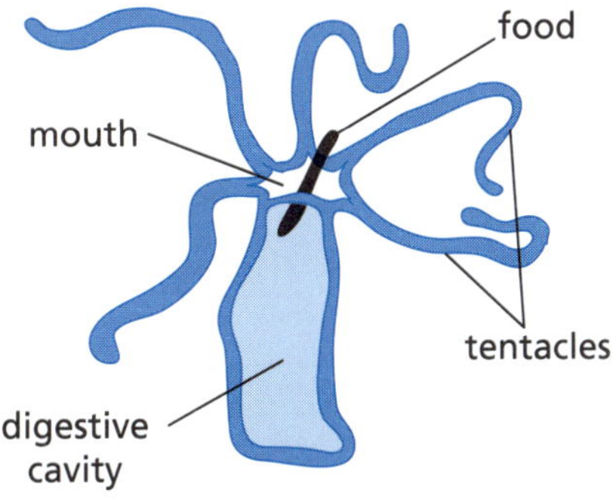

Hint 1: What teeth are missing?
Hint 2: These are the last to be removed.
Hint 3: If too much food entered the intestine at one time, not all of it could be digested.

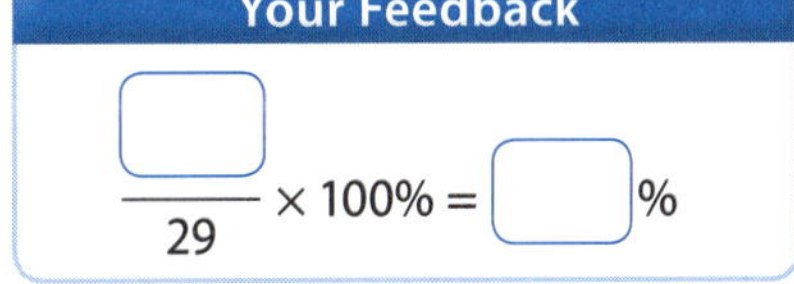

CIRCULATORY AND RESPIRATORY SYSTEMS

Body systems

QUICK REVISION

1 The heart is a powerful, __________ pump. Its role is to pump blood around the body through a variety of different blood __________. Together the heart and the blood vessels are called the __________ system. The heart has a left and right side. On each side there are __________ chambers called the __________ and the ventricle. The left atrium receives __________ blood from the lungs. This blood then passes into the left __________ which contracts and pushes the blood out into a large __________ called the aorta. From there the blood moves through smaller vessels and finally thin __________ in order to deliver oxygen and nutrients to the different organs and cells. The blood also picks up soluble waste and carbon dioxide. The soluble waste is filtered out of the blood by the kidneys. The deoxygenated blood __________ back through veins to the right atrium. This chamber then pumps the blood back to the __________ to be re-oxygenated, and so the cycle continues.

2 If a sample of blood is placed in a centrifuge and spun, the solid components of the blood sediment out. What is left is a pale __________ solution called plasma. This plasma is an important product derived from donated blood. It is __________ into patients who have suffered severe burns in a fire. The plasma replaces lost fluids and blood proteins. The oxygen in our blood is carried in the __________ blood cells. These cells contain red haemoglobin __________ which readily combine with oxygen molecules obtained from the lung. In the __________ some of the carbon dioxide waste is transported by the red cells and the rest is __________ in the blood plasma. Mature red blood cells do not have a __________ and so must be replaced continually after they die. The white blood cells are involved in defending our bodies from attack by __________ microbes. The white cells can move out of the blood capillaries and into the tissues to attack __________-producing organisms. Platelets are fragments of larger blood cells and they are involved in blood __________ formation following an injury to the tissues.

3 Our lungs are a major __________ of the respiratory system. The function of the respiratory system is to transport __________ to the blood and to remove gaseous carbon dioxide __________ from the blood. Oxygen comes into the lungs during inhalation and carbon dioxide is __________ during exhalation. These gases pass up and down your __________ or trachea. This tube is strengthened by rings of __________ to allow it to withstand the pressure changes. The ability to inhale and exhale is coordinated by the brain, __________ and ribs. In order to inhale the chest cavity must be made larger by the __________ cage expanding out and the muscular diaphragm expanding downwards. In order to breathe out, the muscles allow the rib cage to contract to make the __________ cavity smaller. The diaphragm also relaxes upwards. The air we breathe out still contains some of the __________ originally breathed in but it contains more carbon dioxide than before.

Answers **1** muscular; vessels; circulatory; two; atrium; oxygenated; ventricle; artery; capillaries; flows; lungs **2** yellow; transfused; red; molecules; veins; dissolved; nucleus; pathogenic; disease; clot **3** organ; oxygen; waste; removed; windpipe; cartilage; muscles; rib; chest; oxygen

1 In humans, and mammals in general, the body is richly supplied with a network of blood vessels connected to the **heart**, which is a muscular pump. This system of heart and blood vessels is called the **circulatory system**. The heart is a double pump because one side (the right side) pumps blood to and from the lungs, and the other side pumps blood to and from the body. Arteries and veins are the major blood vessels. Arteries have much thicker and more elastic muscular walls compared to veins. The blood pressure is higher in arteries than in veins. Blood which is low in oxygen and rich in carbon dioxide is called deoxygenated blood. It returns from the body via a vein (the **vena cava**) to the heart. The heart pumps this blood to the lungs where gaseous exchange occurs in the fine blood vessels called **capillaries**. Carbon dioxide is released into the lungs and oxygen is absorbed into the blood. This oxygenated blood returns to the heart before it is pumped to the rest of the body. Oxygenated blood leaves the heart via the **aorta**, which is the major **artery** of the body. This artery branches to the head, arms, liver, intestines, kidneys and legs. The intestine and liver contribute nutrients to the blood. The nutrients are carried to the cells of the body where cell respiration occurs. Respiration produces energy to power the body and to keep it warm.

Wastes produced by respiration diffuse back into blood capillaries and the deoxygenated blood is returned via the veins back to the heart. Liquid wastes are filtered from the blood by the kidneys. The following diagrams show the structures of the heart and its place within the circulatory system.

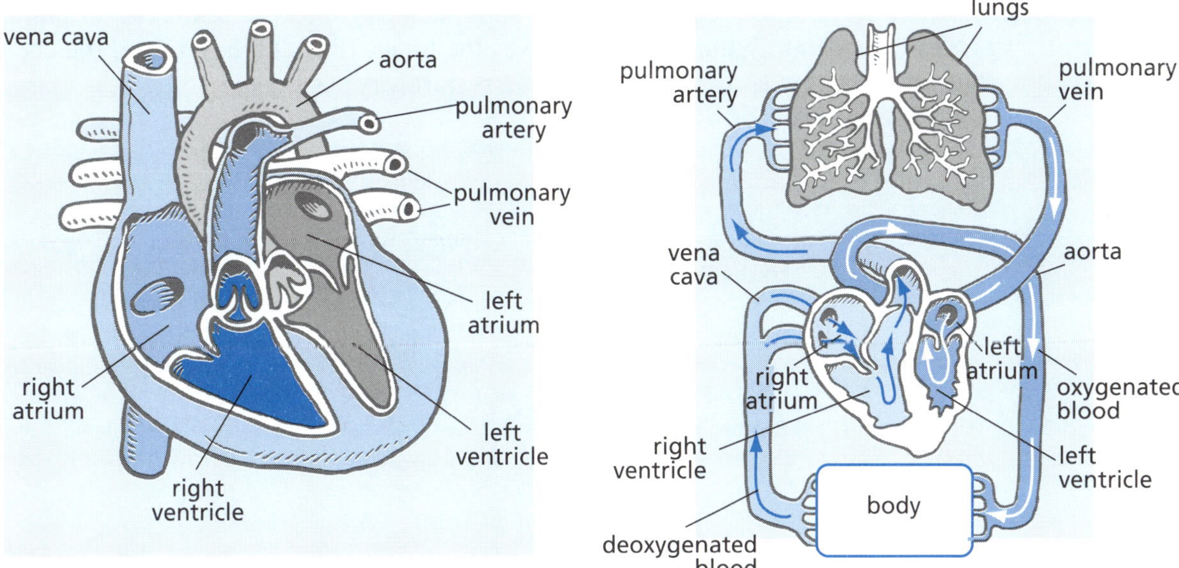

2 Blood is a watery fluid composed of a pale yellow liquid called **plasma** and suspended solids. The plasma contains nutrients, hormones, waste products and various dissolved gases. The plasma makes up 55% of the blood volume. There are three major solid components in the blood.

These are:

- red blood cells (erythrocytes)
- white blood cells (leucocytes)
- platelets.

(cont.)

Red cells are disc-shaped and do not possess a nucleus at maturity. Their active life is around three to four months. Their major function is to transport oxygen from the lungs to the cells of the body. The oxygen binds to the red **haemoglobin** molecules present in the red blood cells. In this way it can be carried around the body. A proportion of carbon dioxide is also transported by red blood cells from the body cells back to the lung. White blood cells possess a nucleus. Their role is to defend the body against disease. They can move from the blood into various tissues where there are sites of infection. Platelets are fragments of large cells. They assist in the process of blood clotting and wound healing and have a short life in the body (up to 10 days).

3 The respiratory system is closely linked to the circulatory system. The respiratory system is involved in bringing oxygen into the body and removing carbon dioxide waste which is produced during cellular respiration. When you inhale, air travels from your nose into the **pharynx** and then into the **trachea** or windpipe. Inhalation is the response to the diaphragm muscle contracting downwards to create a larger space in the chest. The trachea is strengthened with rings of cartilage to ensure it does not collapse during the breathing process. The air then moves into the **bronchi** and subsequently into the smaller **bronchioles** deep in the **lung** tissue. Ultimately, the air reaches small air sacs called **alveoli** clustered at the ends of the bronchioles. Alveoli have a large surface area to ensure rapid exchange of gases. Oxygen diffuses from the alveoli into blood capillaries; carbon dioxide diffuses from the blood capillaries back into the alveoli. When you exhale the diaphragm muscle relaxes upwards, and the gases (now richer in carbon dioxide and poorer in oxygen) retrace this path back to your nose. Exhaling and inhaling occurs by a process called **ventilation**, which involves the lungs, ribs, diaphragm and the chest cavity. The following diagram shows the parts of the respiratory system.

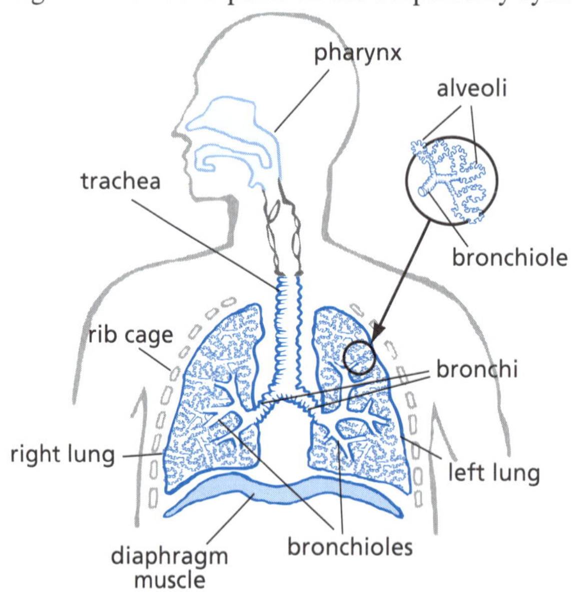

Checklist

Can you:

1 *Identify the purpose of the circulatory system and trace the route of the blood around the body?* ☐
2 *Name the components of the blood and state their function?* ☐
3 *Identify the purpose of the respiratory system and state the purpose of each component?* ☐

REVISION TEST

1 The following diagram shows information about the three major blood vessels. They are labelled X, Y and Z. Use this information to identify the types of blood vessel illustrated. (3 marks)

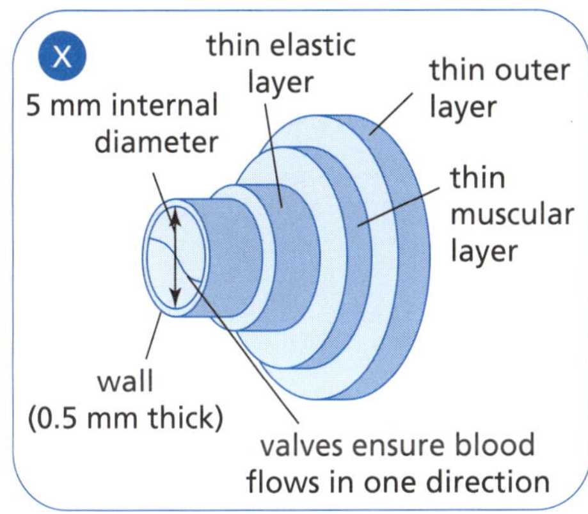

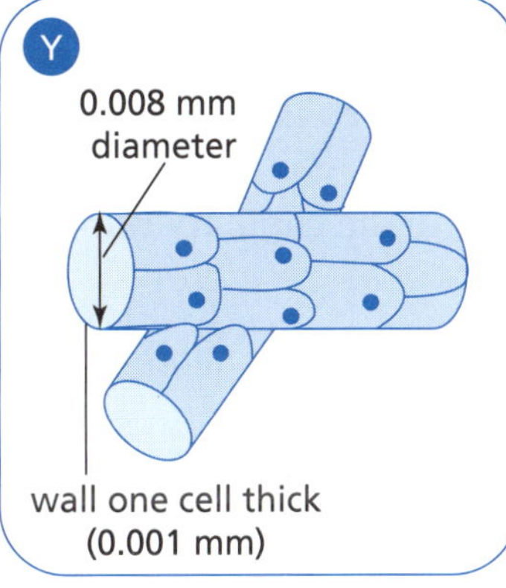

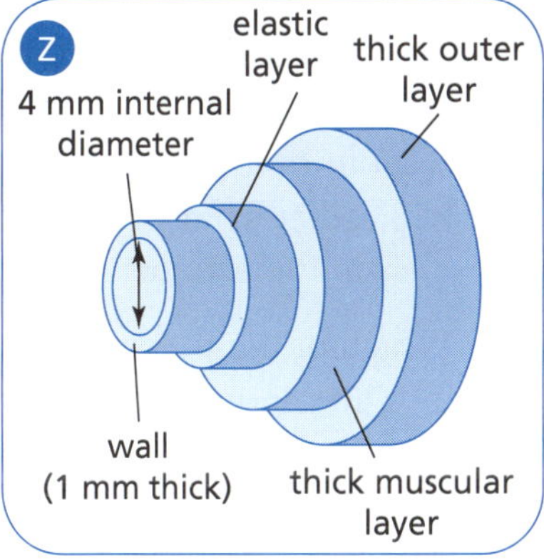

2 Name the vessel of the human circulatory system that transports blood between each of the following.

- **a** heart to the lungs (1 mark)
- **b** heart to the body *Hint 1* (1 mark)
- **c** body back to the heart (1 mark)
- **d** lungs back to the heart (1 mark)

3
- **a** What is the purpose of blood capillaries? *Hint 2* (1 mark)
- **b** Identify the function of white blood cells. (1 mark)

4 The following table shows the typical variation in blood pressure in the heart and blood vessels for a healthy adult. The high pressure reading corresponds to the contraction of the ventricles and the low pressure reading corresponds to the relaxation of the ventricles.

Chamber/vessel	Blood pressure range (mm of mercury) (high–low)
right ventricle	28–0
left ventricle	120–0
pulmonary vein	2–minus 2
capillaries in body tissues	35–8
pulmonary artery	28–10
veins	8–minus 2
main arteries of the body	120–80

- **a** Examine the data. The information is not in a logical order. Arrange the data in the order in which blood flows through the circulatory system, starting with the right ventricle. (1 mark)
- **b** Plot a column graph of the ordered data. (4 marks)
- **c** In which blood vessels is the blood pressure consistently high? (1 mark)
- **d** In which is it consistently low? (1 mark)

(cont.)

e The right ventricle has a thinner muscular wall than the left ventricle. Does this have any effect on the blood pressure in the vessels that lead out of these chambers? Explain. (1 mark)

f Use the data to explain why the compression of veins by nearby body muscles is necessary to move blood through these vessels. (1 mark)

g When you go to the doctor to have your blood pressure taken, the doctor is happy if your readings are round 120/80 (read as 120 over 80). In which vessels is the doctor measuring your blood pressure? (1 mark)

h Predict what would happen to the capillary walls if the blood pressure was as high there as it is in the arteries. (1 mark)

5 Which of the following statements are true and which are false?

a The windpipe is also called the pharynx. (1 mark)

b Alveoli increase the surface area of the lung for gaseous exchange. (1 mark)

c Carbon dioxide diffuses from the alveoli into the blood capillaries. (1 mark)

d Air is exhaled from your lungs when the diaphragm contracts downwards. (1 mark)

e The bronchioles in the lung are strengthened by rings of cartilage. (1 mark)

6 The following diagram shows the inside of the chest.

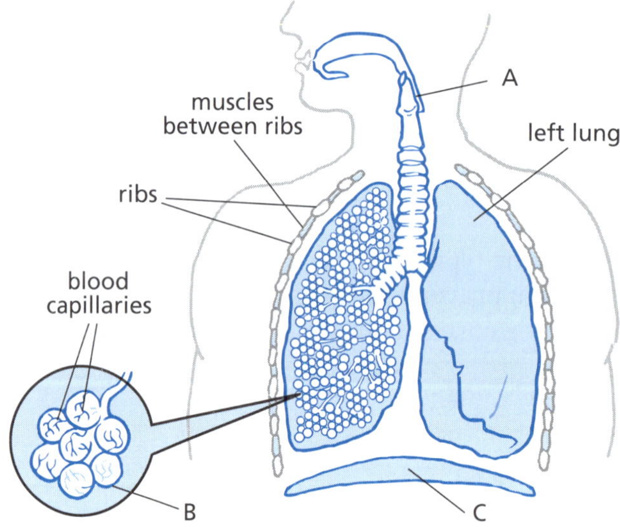

a Name tube A. (1 mark)

b Name the air sacs B. (1 mark)

c Name the muscle C. (1 mark)

d When muscle C moves down, are you breathing in or breathing out? *Hint 3* (1 mark)

Hint 1: This is the largest blood vessel with the thickest muscular wall.
Hint 2: Capillaries are near every cell of your body.
Hint 3: Will your chest cavity increase or decrease if muscle C moves down?

Your Feedback

$\frac{\square}{29} \times 100\% = \square\%$

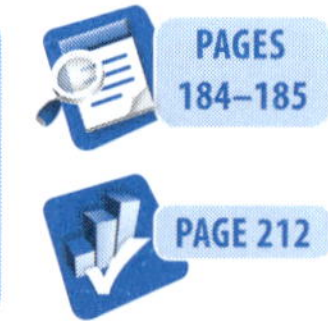

PAGES 184–185
PAGE 212

EXCRETORY AND MUSCULOSKELETAL SYSTEMS

Body systems

QUICK REVISION

1 All organisms take in food and oxygen and __________ wastes. Humans can excrete through their __________ (sweat), their __________ (exhaled moist air containing an increased concentration of carbon dioxide) and through their __________ (urine is temporarily stored in the bladder).

The kidneys are major organs of the human __________ system which removes wastes from the human body. Its excretory activity is affected by __________ that regulate the amount of absorption that occurs.

Humans have __________ kidneys, which are bean-shaped organs located on either side of the lower back. Each is about the size of a fist. Blood enters the kidneys through arteries and leaves through __________. A tube called the __________ carries waste products from each kidney to the urinary __________ for storage and for later release.

The kidneys produce __________, a watery solution of waste products, salts, organic compounds and two important nitrogen compounds: uric acid and urea. Both of these soluble nitrogen products can be __________ to the body and must be removed in the urine.

2 The skeletal system includes all of the __________ in the body and the tissues such as tendons, ligaments and __________ that connect them. The main function of the skeleton is to provide __________ for our bodies. Without a skeleton our body would collapse into a heap. The skeleton is __________ but light and so it also helps protect internal organs and fragile body parts. The brain, eyes, heart, lungs and spinal cord are all protected by your skeleton. The cranium (__________) protects the __________ and eyes, while the ribs protect the __________ and lungs, and __________ in your spinal column protect your spinal cord. As well, bones provide the structure for __________ to attach so that we are able to move. Muscle is attached to bone by tough, inelastic bands called __________.

Bones need regular exercise to remain __________ and healthy. Walking, jogging, running and other __________ activities are important. The skeleton can be strengthened by drinking milk and eating other __________ products. These foods all contain __________, which helps bones harden and become strong.

3 Much of a human's body __________ is in muscles. There are over 630 muscles that move you. They do everything from pumping __________ throughout your body, to helping you lift heavy __________, to running and jumping.

Muscles can't push, only __________. Since muscles often work in pairs, they can pull in different or __________ directions. For example, one muscle, the biceps, will help you raise your __________ arm, while another __________, the triceps, will lower it.

Muscles are all made of the same material, a type of elastic __________ somewhat like the material in a rubber band. Many thousands of small fibres make up each muscle. When muscles receive __________ signals, the cells that make up muscles contract and then __________ back to their original size. The tiny microscopic fibres in these cells compress by __________ in past each other. Muscle cells use chemical energy from __________ to do this.

Answers **1** eliminate (excrete); skin; lungs; kidneys; excretory; hormones; two; veins; ureter; bladder; urine; poisonous
2 bones; cartilage; support; strong; skull; brain; heart; vertebrae; muscles; tendons; strong; physical; dairy; calcium
3 weight (mass); blood; objects; pull; opposite; lower; muscle; tissue (material); nerve; relax; sliding; food

EXCRETORY AND MUSCULOSKELETAL SYSTEMS

Body systems

REVISION SUMMARIES

1 **Excretion** is the process by which waste products of **metabolism** and other non-useful materials are eliminated from an organism. This is mainly carried out by the lungs, kidneys and skin. A major organ for eliminating waste liquids is the **kidney**. Healthy kidneys act like filters to regulate the right amount of wastes and fluids removed. They have a number of vital roles:

- filter the bloodstream to clean it of wastes
- maintain the proper balance of water, salts and chemicals in the body
- produce hormones and enzymes which help to maintain internal water balance
- control blood pressure
- maintain your blood composition and pH levels.

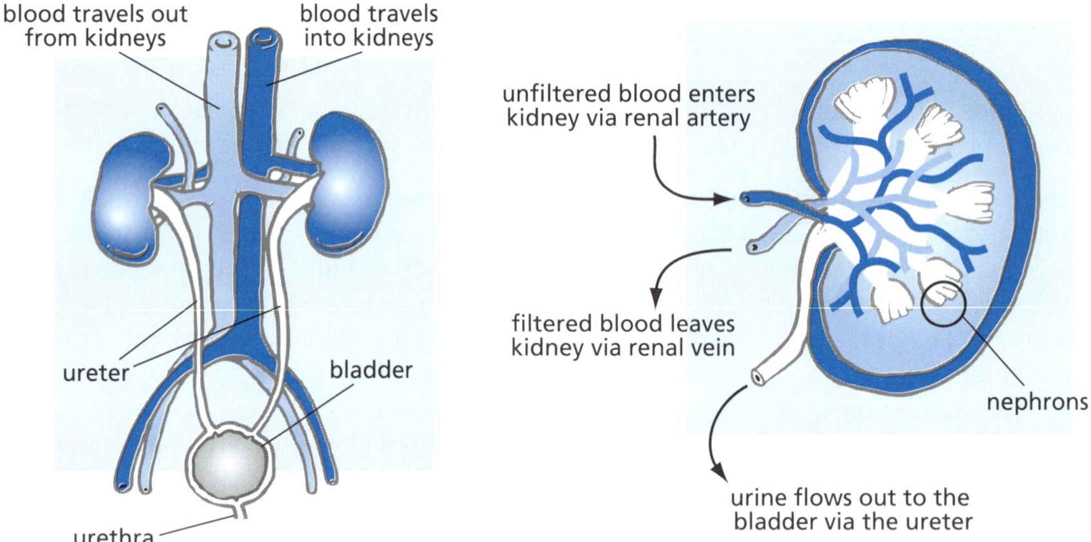

Inside each kidney there are about 1 million tiny filtering units called **nephrons**. As blood passes through the nephron, fluid and waste products are filtered out. Much of the fluid is then returned to the bloodstream, while the waste products are concentrated in any extra fluid as urine. The urine passes out through the ureter, and is stored in the urinary bladder until it is time to urinate.

Defecation is the process by which organisms eliminate solid, semisolid or liquid waste material (faeces) from the digestive tract through the anus. This material was never really absorbed into the body, but rather is the final act of digestion, so it is not called excretion as it is not releasing wastes from metabolism.

2 The **skeletal system** gives us our shapes and support. But it also protects soft body parts such as the brain, lungs and heart. As well, we use our skeleton for movement. By contracting muscles attached to our bones we are able to move about, play games and even master fine skills such as playing musical instruments.

Joints allow us to move. They provide flexible connections between bones. There are four types of joints:

- **pivot** (e.g. the neck allows for the bones to rotate, but not all the way around)
- **ball and socket** (e.g. the shoulder enables bones to move 360°)

- **hinge** (e.g. the knee and elbow allow bones to move backwards or forwards)
- **ellipsoidal** (e.g. the wrist and ankle allow bones to slide one over the other).

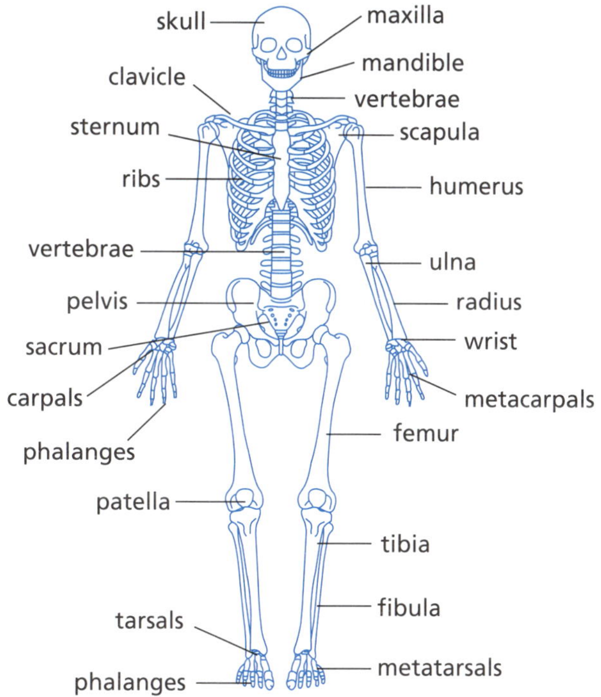

In the body, bones are alive with living cells which help them grow and repair themselves, and have their own nerves and blood vessels. They store body minerals like calcium. In the jelly-like **bone marrow** of some bones, new red and white blood cells are constantly being produced. A typical bone consists of an outer layer of hard, dense, tough and strong compact bone. Inside this is a lighter and slightly flexible layer of spongy bone. The following diagram shows two different cross-sections through a large bone.

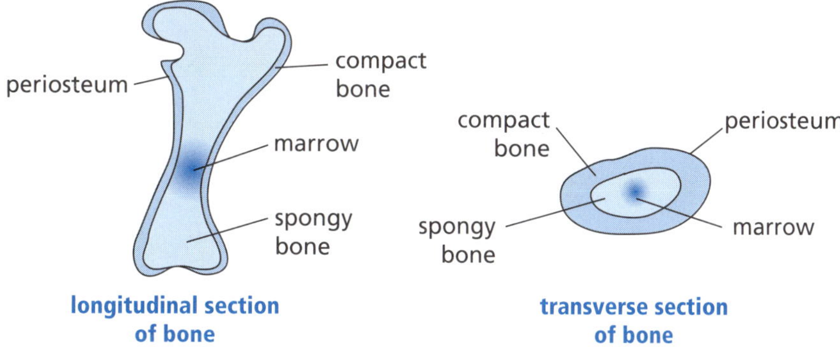

longitudinal section of bone

transverse section of bone

3 **Muscles** have the ability to contract, allowing for body movements and functions. Muscles are mainly composed of alternating rows of **protein** filaments. When the muscle is relaxed, those rows are least overlapped. When a nervous impulse commands the muscle to contract, these rows overlap and shorten the muscle length.

(cont.)

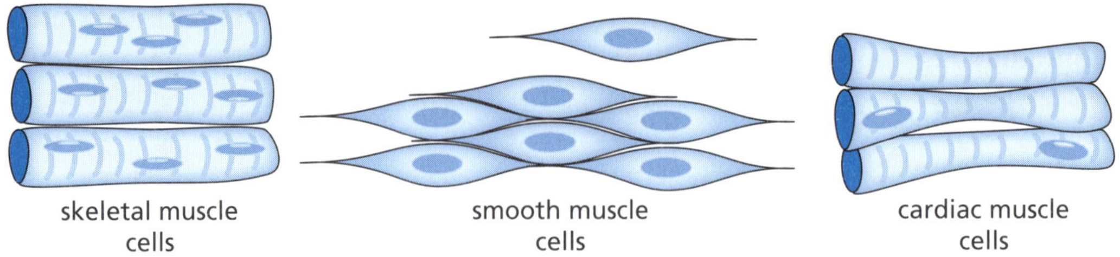

The muscular system consist of three different types of muscle tissues.

- **Skeletal.** These make up around 40% of an adult's body weight and are connected to bone (skeleton). These muscles consist of long muscle fibres. The nervous system controls their contraction. Many of the skeletal muscle contractions are automatic. We can walk automatically without needing to think of the action. However, we still can control the actions of skeletal muscles.
- **Cardiac.** This is the muscle of the heart itself. Like skeletal muscles, the cardiac muscle is striated (striped) but it is different to skeletal muscles because it is not attached to a bone.
- **Smooth.** Many internal organs contain smooth muscles. They are found in the urinary bladder, digestive tract, gall bladder, arteries and veins. Smooth muscles are controlled by the nervous system and hormones. Like the heart muscle, we cannot consciously control smooth muscles. This is why they are often called involuntary muscles.

Checklist

Can you:

1 *Describe the function and working of the human excretory system?* ☐

2 *Describe the function and working of the human skeletal system?* ☐

3 *Describe the function and working of the human muscular system?* ☐

EXCRETORY AND MUSCULOSKELETAL SYSTEMS

Body systems

REVISION TEST

1 The amount of urine produced by a person each day depends on a number of factors. Give at least three of these factors. *Hint 1* (3 marks)

2 At birth, babies have about 300 bones. Adults, on the other hand, have about 206 bones. Suggest what might happen to the other bones. *Hint 2* (1 mark)

3 The body's total blood supply circulates through the kidneys about 12 times per hour.

a If a person has about 5 L of blood in his system, how much blood passes through the kidneys:
 i each hour? ii each day? (1 mark each)

b If, on average, a person produces 1.5 L of urine each day, what percentage is this of the total blood filtered each day? *Hint 3* (1 mark)

4

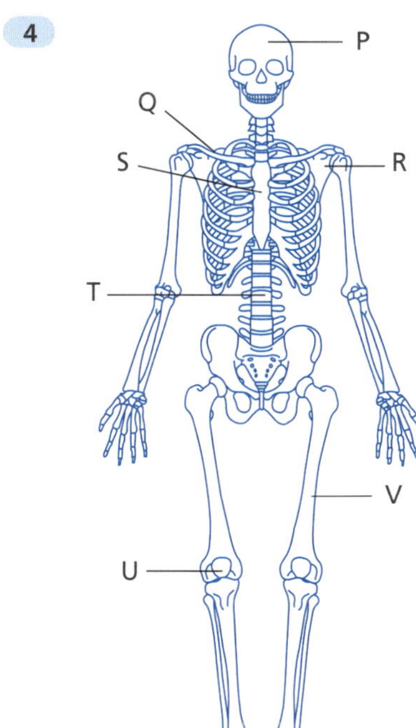

This skeleton shows the bones of the body. Label each statement below using an appropriate letter, and give the scientific name for the bone referred to.

a The thigh bone is the largest bone in the body. (2 marks)

b Known as the kneecap, it is a thick circular-triangular bone which articulates with the femur. (2 marks)

c The collarbone is a long bone of short length that serves as a strut between the scapula and the sternum. (2 marks)

d The skull without a mandible has functions that include protecting the brain, fixing the distance between the eyes allowing stereoscopic vision and fixing the position of the ears. (2 marks)

e Sometimes called a backbone, it is not a single bone but consists of many small bones. (2 marks)

f The shoulder blade is the bone that connects the humerus with the clavicle. (2 marks)

g The breastbone is a long flat bony plate located in the centre of the thorax (chest) and connects the rib bones through cartilage, helping to protect the lungs, heart and major blood vessels from physical trauma. (2 marks)

5 There are two types of muscles in the system: involuntary muscles and voluntary muscles.

a Describe the difference between voluntary and involuntary muscles. (2 marks)

b Give two examples of each type. (4 marks)

c What causes these muscles to move? (1 mark)

6 a Name the bones shown in the diagram. (3 marks)

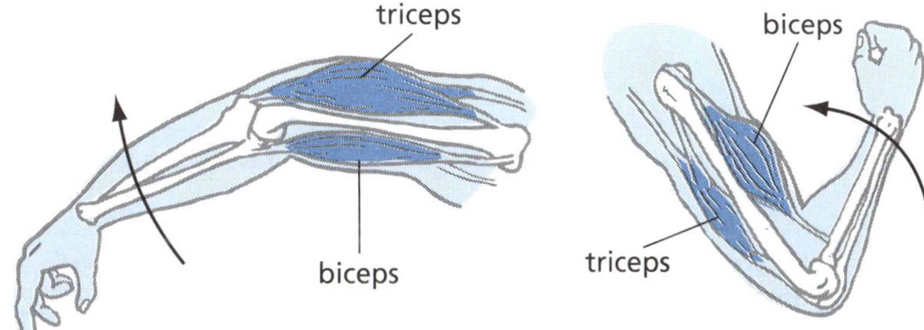

(cont.)

b What type of joint is shown? (1 mark)

c How are the muscles connected to the bones? (1 mark)

d Describe how the biceps and triceps muscles can raise and lower your forearm. (4 marks)

7 a To what system does the following diagram belong? (1 mark)

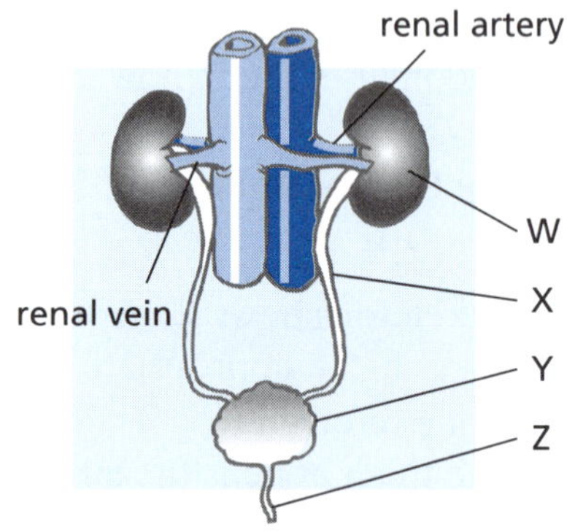

b Name the parts labelled W, X, Y and Z. (4 marks)

c Describe the main function of the kidneys. (3 marks)

8 Certain kinds of remains, such as bones, can provide information about the physical characteristics of ancient people, and from incomplete skeletons found at crime scenes. For example, given a femur, the approximate height of the individual can be inferred.

Male: Height (cm) = $1.880 \times$ femur length (cm) + 81.31

Female: Height (cm) = $1.945 \times$ femur length (cm) + 72.84

Suppose a femur measuring 45.5 cm is found. Estimate the height of the individual if that person is

a male. (1 mark)

b female. (1 mark)

Hint 1: Think of what factors in your day-to-day activities may contribute to your going to the toilet more or less often.
Hint 2: They do not disappear or dissolve away.
Hint 3: To calculate the percentage: (volume of urine each day)/(total blood volume filtered each day) × 100

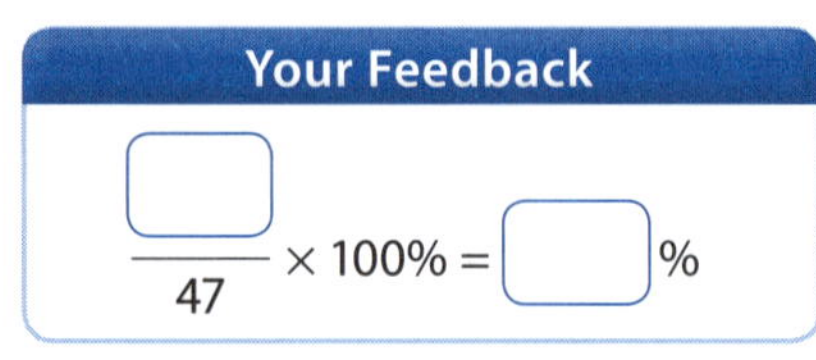

ASEXUAL AND SEXUAL REPRODUCTION

Body systems

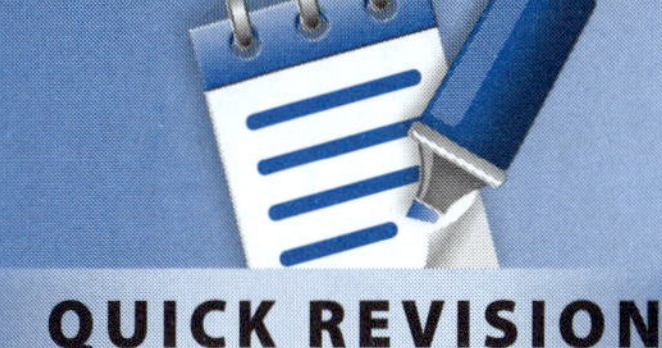

QUICK REVISION

1 Reproduction is the ________ of new individuals from previously existing ________. Sexual reproduction needs two parents. But in ________ reproduction only one parent is needed. These ________ are produced by mitosis. Since there is only one parent, there is no fusion of ________ (sperm and egg) and hence no mixing of genetic information. As a result, the offspring are genetically ________ to the parent and to each other. In other words, they are clones.

Many plants can reproduce ________. Some plants develop underground food storage organs which later form the following year's plants. For example, potato and daffodil plants do this. Other plants produce side branches where small ________ develop on them. Other plants, such as strawberries, produce runners with plantlets on them. Many plants around the home can be propagated asexually. Cuttings (tip and stem, cane, whole leaf, leaf section, leaf bud), division, runners and air layering are all ways of asexually ________ plants. Asexual reproduction in animals does occur, such as in sea anemones and starfish, for example, but it is less common than sexual reproduction.

2 Humans, like many other advanced animals, cannot reproduce ________. In sexual reproduction a new individual is produced following the ________ together (union) of two gametes (sex cells). These ________ cells differ in structure and are contributed by different parents. The gametes (male ________) need to be mobile to be able to meet and unite, and provide food (in the female ________ or ovum) to nourish the developing embryo.

In males, sperm production occurs in the ________. Sperm is produced by the process of meiosis where the number of ________ is halved. This sperm then needs to be delivered to the reproductive tract of the female. In females, eggs are formed in the ________. Eggs are also produced by meiosis where the number of chromosomes is halved. Further, the female must be equipped to receive sperm from the male.

Fertilisation is the process of ________ together the sperm and egg so the two half-number sets of chromosomes become one complete set again. The fertilised egg (________) now undergoes mitosis forming a mass of cells. Approximately one week after fertilisation, this bundle of cells embeds itself in the thickened wall of the ________ and pregnancy is established.

3 In flowering plants sexual reproduction centres on the ________. Within a flower, there are usually structures that produce both male ________ and female gametes. Eggs develop inside the ovary. Each ________ (male part) contains pollen sacs. Pollination is the process by which ________ is transferred to the stigma of plants, and so allow fertilisation and sexual ________ to occur. Many plants favour cross-pollination, so pollen must be transferred to the ________ of another plant if sexual reproduction is to take place. Some flowers rely on the wind to carry pollen grains, while others rely on ________. Self-pollination may also occur—this is where the pollen is transferred to the stigma of the same or another flower but on the ________ plant. Self-pollination is more reliable, especially if the nearest plant is not very close. If the pollen grain lands on a stigma, a pollen ________ will grow so that eventually the egg cell, hidden away in the ________, can be fertilised. A tube emerges from the grain, and grows down through the style to the ovary.

Answers **1** creation (formation); individuals; asexual; offspring; gametes; identical; asexually; plantlets; reproducing **2** asexually; fusing; sex; sperm; egg; testes; chromosomes; ovaries; joining (uniting); zygote; uterus **3** flower; gametes; anther; pollen; reproduction; stigma; insects (animals); same; tube; ovary

ASEXUAL AND SEXUAL REPRODUCTION

Body systems

REVISION SUMMARIES

1 **Asexual reproduction** is a form of reproduction where offspring arise from a single parent, and inherit the genes (characteristics) of that parent only. Advantages of asexual reproduction include no need to search for a mate, and that every asexual organism can reproduce on its own. This can therefore lead to a rapid population build-up. However, every new organism produced by asexual reproduction is genetically identical to the parent; it is a clone.

Since only identical individuals (**clones**) are produced in asexual reproduction, there is no variation. This is of benefit in agriculture and horticulture where standardised (uniform) production is needed. However, in the wild, an asexual population that cannot adapt to a changing environment, or evolve defences against some new disease, is at risk of extinction. Hence many asexually reproducing organisms are capable of reproducing sexually as well.

Asexual reproduction is very common in microorganisms. Bacteria, for example, reproduce through a process called **binary fission**. During reproduction, the bacteria duplicate the genetic information on the chromosome and then split into two as shown in the following diagram. But there are also many plants that use asexual reproduction naturally.

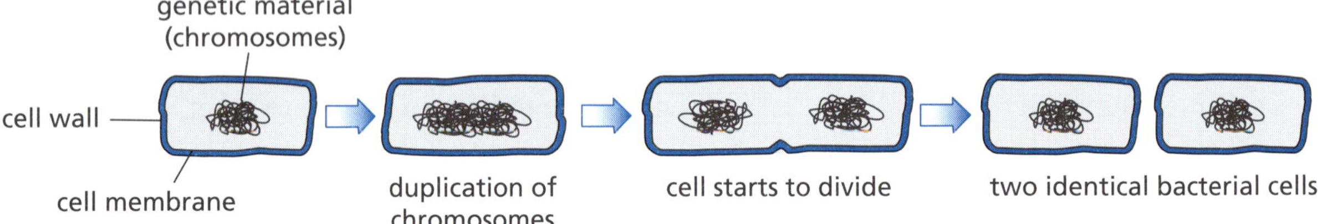

Hydra is a simple animal that lives in freshwater ponds and streams. An adult hydra will develop a swelling (bud) on the side of its body. Eventually this daughter bud grows tentacles and starts to feed itself by catching small water animals. When fully formed, it detaches from the parent as shown in the diagram.

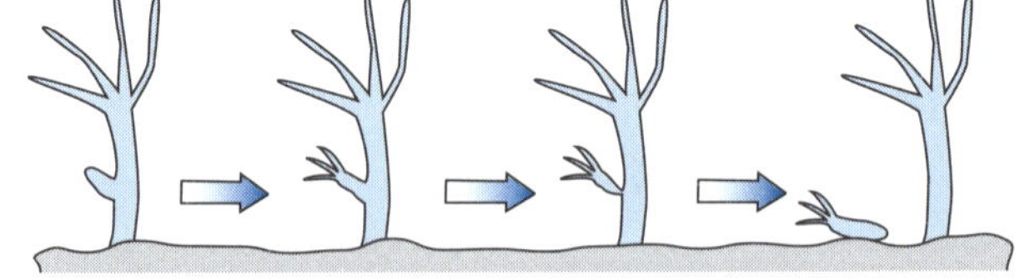

While asexual reproduction occurs in plants and simple animals, scientists have discovered around 70 species of vertebrates that propagate asexually. For example, female whiptail lizards (*Aspidoscelis* genus) from Mexico and south-western United States manage to produce healthy female offspring without requiring male fertilisation.

2 Asexual reproduction is the production of genetically identical offspring from a single parent. Sexual reproduction, on the other hand, is the creation of a new organism by combining the genetic material of two parent organisms.

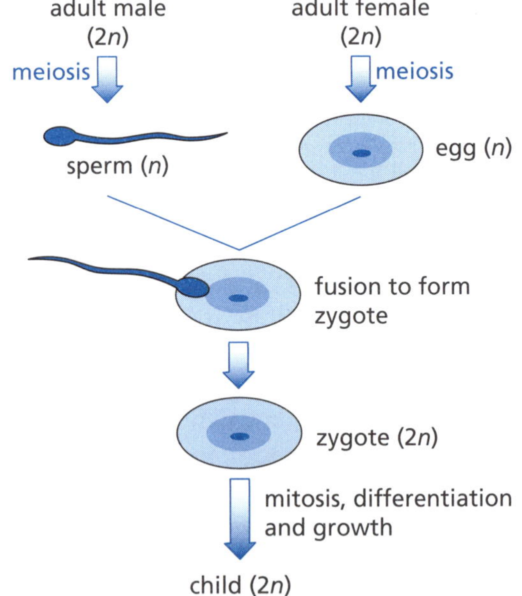

There are two main processes during sexual reproduction:

- **meiosis**, involving halving of the number of chromosomes
- **fertilisation**, involving fusing two gametes and restoring the original number of chromosomes.

Humans have 23 pairs ($2n = 46$) of chromosomes in each body cell. A sperm carries $n = 23$ and an egg carries $n = 23$, so when fertilisation occurs the $2n$ number is restored. During meiosis, the chromosomes of each pair usually exchange information (genes) creating variation in the new organism.

3 Unlike animals, plants can't move around, and so cannot seek out sexual partners to reproduce. Flowering plants are the dominant plant form on land and can reproduce by sexual and asexual means. The **reproductive organs** of these plants are simply called flowers. Flowers are both male and female, although there are some plants where the sexes are separate. Sexual reproduction involves producing male and female gametes, then transferring the male gametes to the female ovules in a process called **pollination**.

Pollination is the process of transferring the sticky pollen from a stamen to the stigma. Usually plants rely on animals or the wind to pollinate them. Plants that are pollinated by animals such as insects often have brightly coloured flowers with a strong scent to attract the animal pollinators. They also usually provide nectar (food) for these insects. **Wind-pollinated** flowers often have long stamens and pistils. The wind picks up pollen from one plant and blows it onto another. Since these plants do not need to attract **animal pollinators**, they can be unscented and poorly coloured, with small or no petals. The parts of a flower are shown in the following diagram.

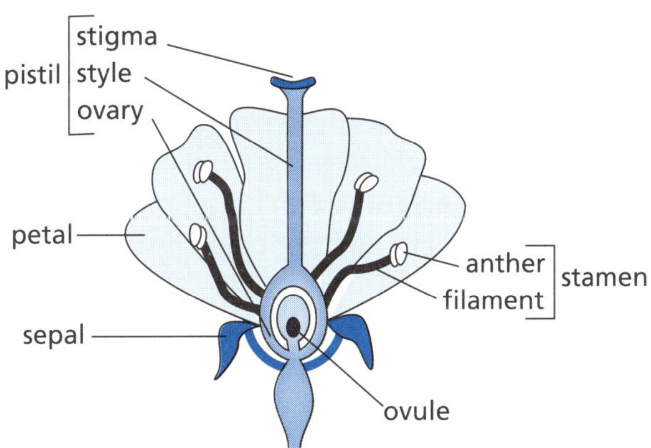

After pollination, fertilisation occurs and the ovules grow into seeds within a fruit. When the seeds are ready for dispersal, the fruit ripens freeing the seeds allowing them to germinate and grow into the next generation.

Checklist

Can you:

1 *Describe the function of asexual reproduction?* ☐

2 *Describe sexual reproduction in humans?* ☐

3 *Describe sexual reproduction in flowering plants?* ☐

ASEXUAL AND SEXUAL REPRODUCTION

Body systems

REVISION TEST

1 The following diagram shows binary fission in the single-celled organism called an amoeba.

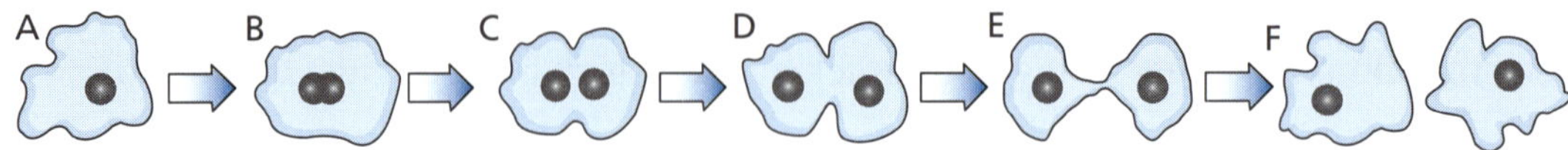

a Is this process sexual or asexual? How do you know? (2 marks)

b Write captions for A to F describing how this process occurs. (6 marks)

2 Air layering can be used to asexually propagate many plants as shown in the following diagram. Describe the steps in this process. (4 marks)

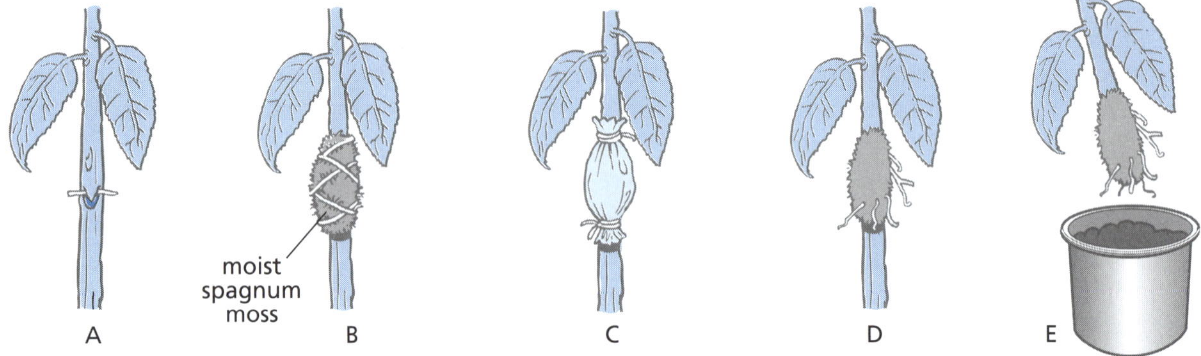

3 Describe three advantages and three disadvantages of asexual reproduction. (6 marks)

4 Two types of pollination are self-pollination and cross-pollination.

a Define each type. (2 marks)

b What is a potential drawback of self-pollination? *Hint 1* (2 marks)

c Cross-pollination has been described as less reliable and more wasteful than self-pollination. Explain. (2 marks)

d Plants that require animals to transfer pollen have flowers that are colourful, have lots of scent, produce sugary juices (nectar) or have appealing shapes and patterns. Why is this? *Hint 2* (2 marks)

e Comment on this statement:
When plants are pollinated by insects, it seems like some kind of agreement was entered into whereby the plant will provide the insect with something, if the insect will pollinate the plant's flowers in return. (1 mark)

5 The following diagram shows parts of a flower. Use the diagram to complete the table below. (14 marks)

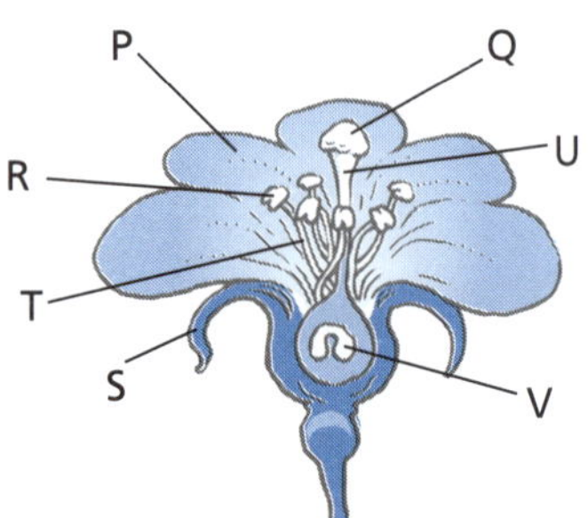

Letter	Name	Function
		contains egg cells
		attracts insects and other pollinators
		traps pollen
		makes the pollen
		used to protect the growing flower bud
		pollen travels through this part of the pistil
		part of the stamen providing support

6 The following diagrams show the human male and female reproductive systems.

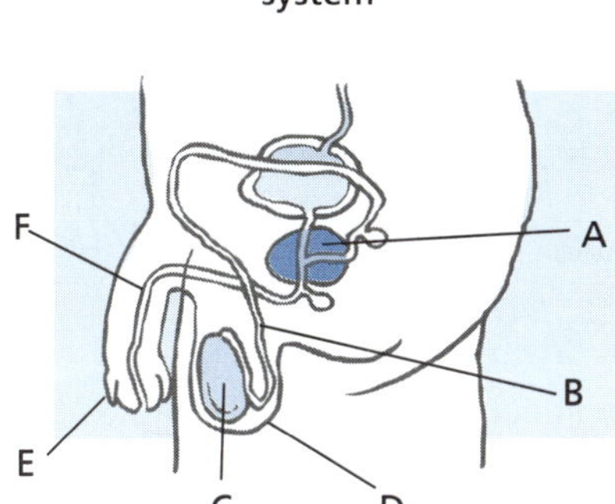

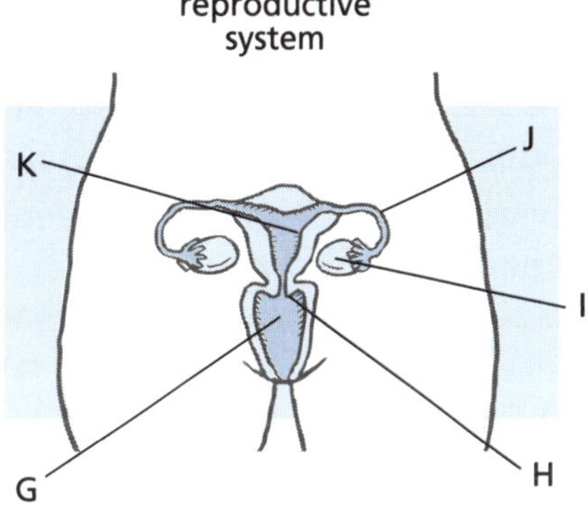

Use the descriptions below to identify the letters on the diagram. *Hint 3* (11 marks)

Male reproductive system		
Letter	**Name**	**Function**
	penis	the organ used to deposit semen into the vagina of the female during intercourse
	urethra	a tube running through the length of the penis that carries semen; can also discharge urine from the bladder
	scrotum	the external sac holding the testes
	testes	site of sperm production; located within the scrotum
	vas deferens	also called the sperm duct, it is a tube carrying sperm away from the testes
	seminal vesicles	one of several glands producing protective and nutrient fluids for the sperm; feeds into the sperm duct

Female reproductive system		
Letter	**Name**	**Function**
	ovaries	site of egg production, located one on each side
	vagina	canal into which semen is deposited and the birth canal through which the baby is born
	fallopian tubes	also known as oviducts, the tubes through which mature eggs pass following ovulation
	uterus	also known as the womb, this is a muscular organ where the fertilised egg will implant and grow to form the baby
	cervix	a narrow entryway between the vagina and uterus; muscles here are flexible expanding to allow baby to pass during birth

Hint 1: This question is effectively asking for a disadvantage.
Hint 2: Over the years flower breeders have developed plants with larger, more colourful and greater variety and more pleasantly scented flowers. This extreme is appealing to us (humans).
Hint 3: The description and function is enough to determine which label is correct.

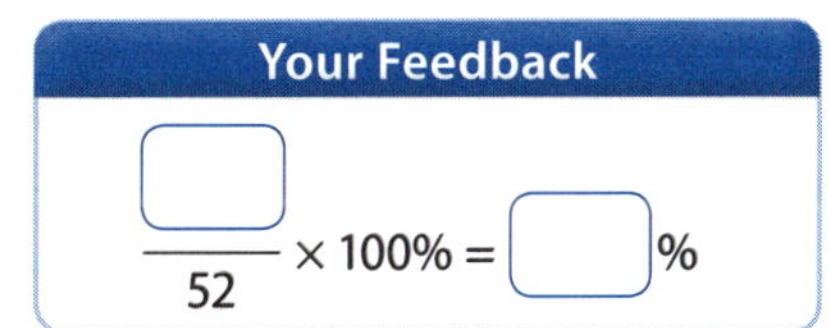

PAGES 185–187
PAGE 212

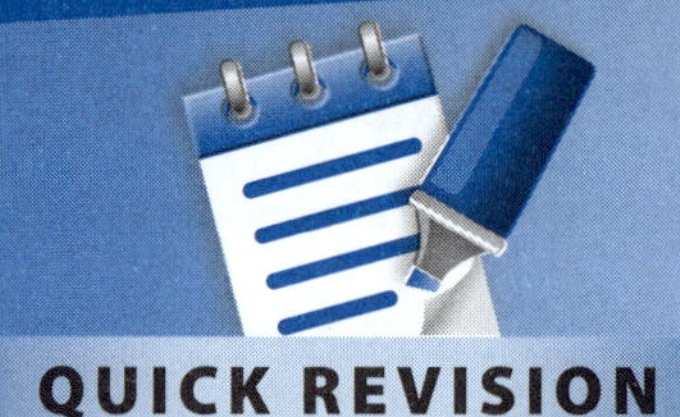

1 Cloning may appear to be an __________ process, but many plants naturally clone themselves to __________. This diagram shows strawberry runners, which are horizontal __________ growing above the ground producing a new (__________) plant.

mother plant
daughter plants
crown
runner

People may want to clone a plant deliberately. This could be because that plant shows __________ characteristics that farmers and agriculturalists desire. Cloning of plants has many important commercial implications. It allows a successful variety of a plant to be produced commercially and __________, on a massive scale in a short space of __________.

The easiest way to clone a plant requires taking a __________. A branch from the parent plant is cut off, and its lower leaves removed. The stem is planted in damp compost, sometimes with the aid of plant hormones to __________ new roots to develop. After a few weeks, new roots have grown producing a new, but __________, plant.

2 As scientists and doctors are learning more about how human reproduction works, they are in a better position to assist people __________ when they wish to have babies and to ensure those children are born healthy. For example, in the past many women who didn't want to have children would undergo an __________. Such procedures could be painful and dangerous. With the invention of the __________ pill, women now have more control over whether or not they get pregnant. On the other hand, there are now drugs available to __________ reproductive fertility. For instance, these can stimulate the release of an egg from the ovary so it can be fertilised. Parents now also have the ability to check on the health of the developing __________ as it is growing inside the womb. If there is a problem, or some genetic defect, parents can be given the __________ of whether or not to abort.

These choices give rise to many bioethical issues, and people have different views on them. Some may argue that if a child is to be born deformed, or with some __________ condition that won't allow it to live for long with any quality of life, is it fair to bring it into the world and have both the child and parents suffer. As more discoveries are made, more questions will arise with no easy __________.

3 In the past when an important part of the body became __________, the patient often died. People with faulty hearts, for example, would die from __________ disease. Since the first heart transplant in 1967, transplanting hearts from brain dead __________ has become fairly common. Other __________ and tissues can now be transplanted too. However, as the body does not recognise these transplanted organs as __________, but rather as non-self, it thinks it is some sort of __________ and the body's natural defence (the __________ system), kicks in to try and get rid of it. This is why transplant patients need to take anti-rejection __________ for the rest of their lives.

Answers **1** artificial; reproduce; stems; cloned; favourable; cheaply; time; cutting; encourage (promote); identical (cloned) **2** control (decide); abortion; contraceptive; enhance (promote, assist); foetus; option; genetic (inherited); answers **3** diseased (sick, damaged); heart; donors; organs; self; infection; immune; medication (medicines, drugs)

1 In biology, **cloning** is an asexual reproduction process that involves producing similar populations of genetically identical individuals. One way to do this is taking a cutting and then grafting it as shown in the diagram.

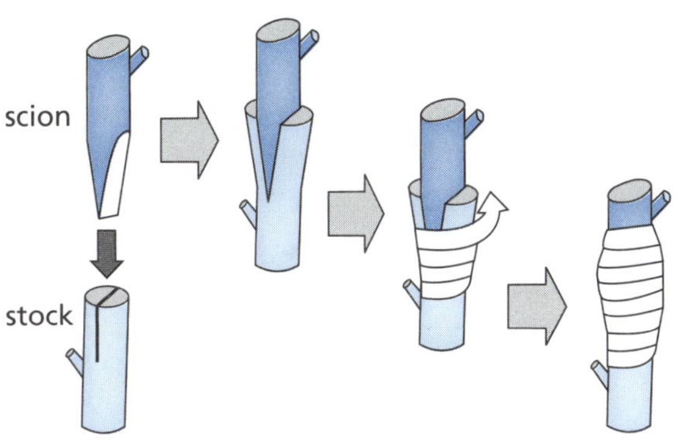

Another way of cloning plants is by **tissue culture**, which works not with cuttings but with tiny pieces from the parent plant as shown in the diagram below. A variety of plant species can be conveniently propagated by cell, tissue or organ culture. This is often called micropropagation. The major benefits include:

- rapid multiplication of superior and consistent (uniform) clones
- production of many disease-free plants
- multiplication of sexually derived sterile hybrids, that is, plants which are unable to reproduce by sexual means.

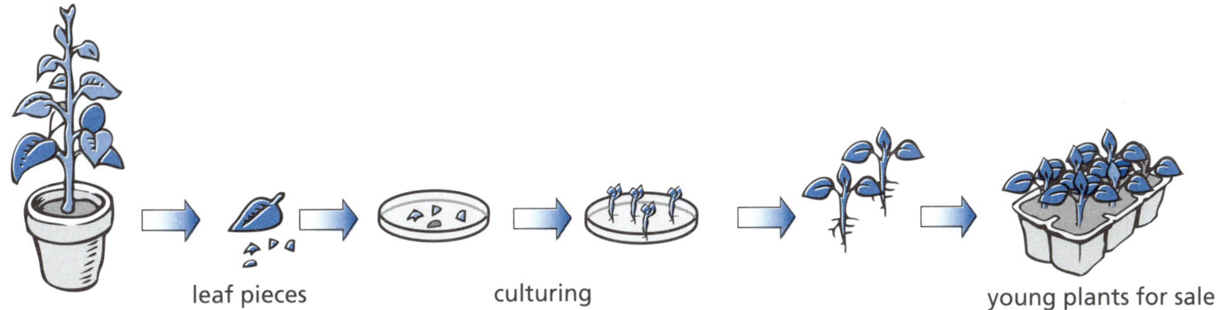

But cloning also occurs naturally in nature. Single-celled organisms like yeast, germs and protozoa make cells exactly like themselves through asexual reproduction. Plants can also produce other plants asexually through **vegetative propagation**. This process is where a stem or root that is planted makes an exact replica of itself. Farmers use this technique to make more plants, fruits and vegetables with desirable qualities.

2 Reproductive technology includes all present and future uses of technology for both human and animal reproduction. This includes assisted reproductive technology and contraception. The following are some examples of reproductive technology.

- **Contraception.** This enables people to control their fertility. Besides the contraceptive pill, there are barrier methods such as condoms and diaphragms. These methods range from daily contraception options to non-daily or long-term reversible contraception.
- **Artificial insemination (AI).** Some males are unable to produce sperm, the sperm is not very motile or he carries a genetic disorder. It can also be used where the female does not have a sexual partner. AI is the deliberate introduction of semen into a female's vagina in order to get her pregnant. This process has long been used in breeding dairy cattle and pigs and in passing on desirable traits, and is now adapted for humans.
- **In-vitro fertilisation (IVF).** In this process the eggs are fertilised with sperm in a fluid medium outside the woman's body. It is a major treatment for infertility. Once the eggs are successfully fertilised, one or several are then transferred to the patient's uterus.

(cont.)

- **Somatic cell nuclear transfer.** Animal or human embryos are created for research purposes. This can assist in the study of human development and possibly treat disease.
- **Reproductive cloning.** This involves transferring a nucleus from a donor adult cell to an egg that has no nucleus. This has allowed scientists to clone animals. Since cloning the first sheep in 1996, scientists have been able to clone dozens of different animals.
- **Cryopreservation.** This process involved cooling cells or whole tissues to very low temperatures, effectively stopping any biological activity. This is especially useful for men undergoing cancer treatments, which could render them sterile. This way they can preserve sperm for possible use later. It is also possible for a woman's eggs to be extracted, frozen and stored.

3 Medicine has progressed to a point where organ transplants have become almost routine in many large hospitals. **Transplantation** is the process of transferring an organ from one person's body to another person, or from one site to another on the person's own body. Organs that can now be transplanted include the heart, lungs, kidneys, intestines and the pancreas, with the kidney being the most commonly transplanted organ. Tissues can also be transplanted. These include bones, tendons, cornea (from the eye), skin, heart valves and veins.

In a **kidney transplant**, for instance, sources of donated kidneys could be living people (such as relatives), since people can live quite well with only one kidney. Another source is brain dead or cardiac (heart) death donors where the kidneys are functioning well. The more closely matched the tissues are, the less chance there is of rejection. However, as medication to prevent rejection is very effective, donors do not need to be genetically similar to the recipients. Nevertheless, immunosuppressant drugs are used to **suppress** the immune system from rejecting the donor kidney and these medicines must be taken for the rest of the patient's life.

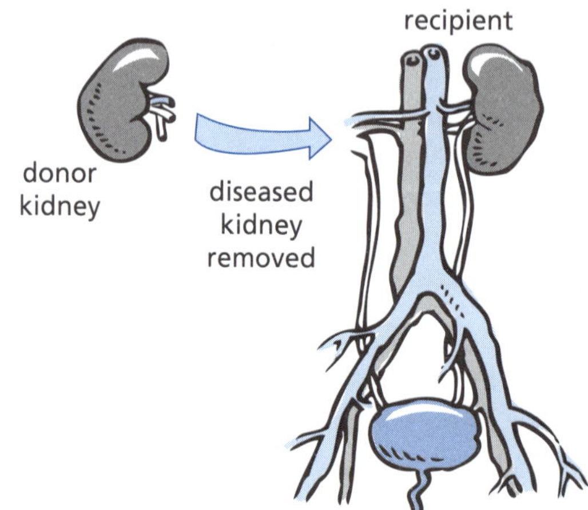

Checklist

Can you:

1 *Recognise cloning and identify different techniques used to clone?* ☐
2 *Define 'reproductive technology' and give several examples?* ☐
3 *Identify what organ transplants are and some difficulties associated with it?* ☐

BIOTECHNOLOGY

REVISION TEST

1 **a** What is shown in the process in the following diagram? (1 mark)

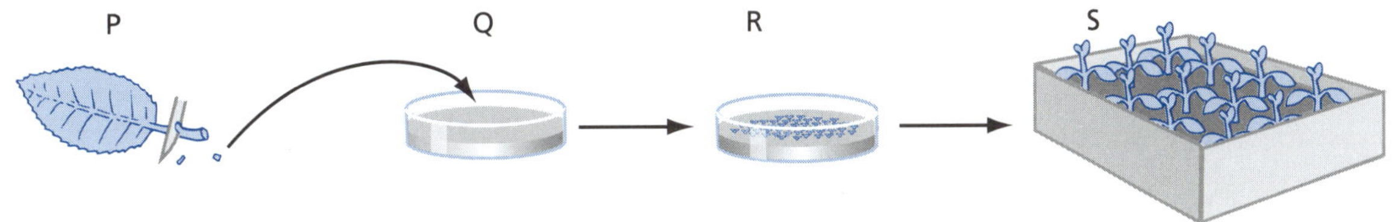

b Write captions to replace letters P, Q, R and S. *Hint 1* (4 marks)

2 Scientists have experimented with cloning animals.

a Suggest a commercial reason for a farmer wanting a cloned animal. (1 mark)

b Dolly the sheep was the first mammal to be cloned using adult cell cloning. She was born in the United Kingdom in 1996 but died in 2003.
The following diagram shows the steps in how she was produced. At the end of this process, Dolly is born and is genetically identical to the donor sheep.

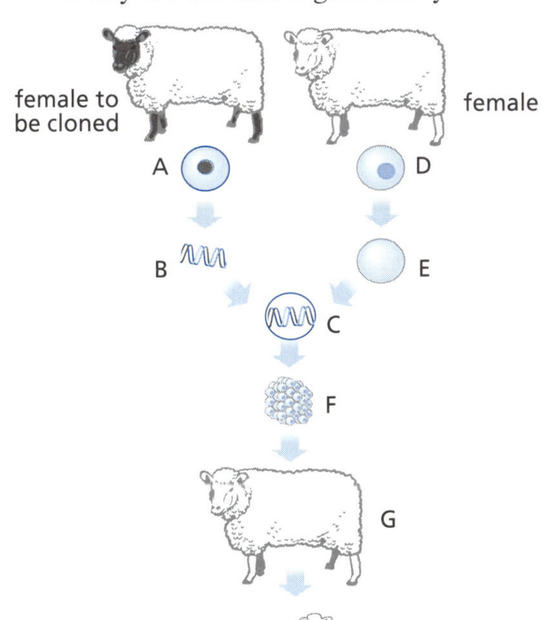

Match each letter in the diagram to a statement listed below. *Hint 2* (7 marks)

i An egg cell is removed from the ovary of an adult female sheep, one that will not be cloned.

ii The nucleus, which contains the genetic code, is removed from the body cell and readied for use.

iii The new fused cell begins to develop normally, using genetic information (code) in the donated nucleus.

iv The nucleus from the donor sheep is inserted into the empty egg cell.

v The nucleus is removed from this egg cell and discarded.

vi The cells begin to divide but before they become specialised, the embryo is implanted into the uterus of a foster mother sheep.

vii A body cell is taken from the sheep to be cloned.

3 List three advantages of tissue culture. (3 marks)

4 Give three examples of reproductive technologies. (3 marks)

5 About 1% of lizards can reproduce by spontaneously ovulating and then cloning themselves to produce offspring with the same genetic blueprint.

a Which sex are lizards that reproduce by spontaneously ovulating: male or female or either? (1 mark)

b What is a potential problem with lizards cloning themselves? *Hint 3* (2 marks)

(cont.)

6 The following diagram shows an example of a reproductive technology.

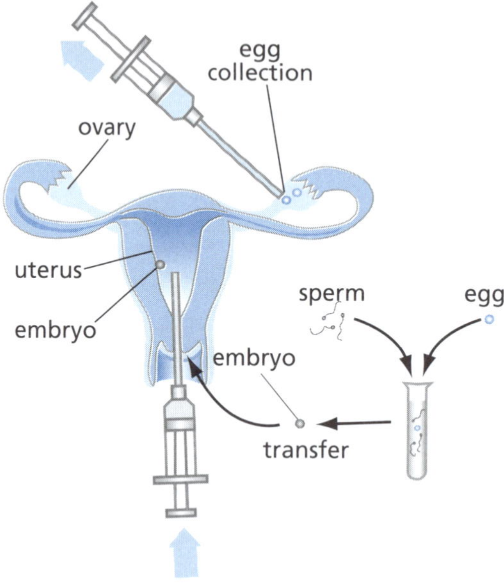

a Which reproductive technology is shown? (1 mark)

b Describe what the diagram is showing. *Hint 4* (4 marks)

7 Kidney transplant patients need to take immunosuppressant drugs.

a What are immunosuppressant drugs, and why are they taken? (2 marks)

b Infections are likely with kidney transplant patients. The sector graph shows the most common infective agents. Which is the most common infective agent in kidney transplant patients? (1 mark)

Common infective agents for kidney transplant patients

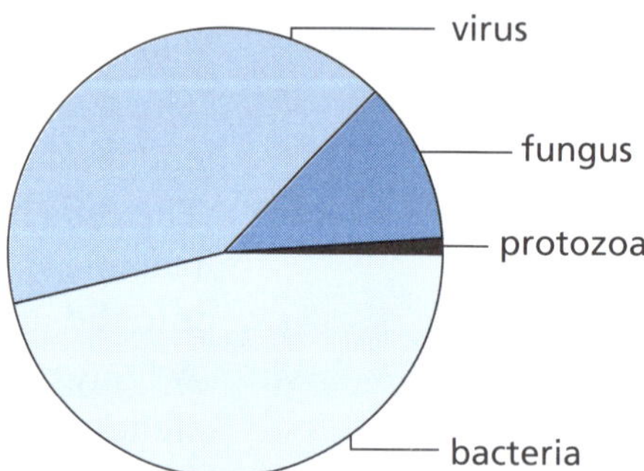

c Why are these agents sometimes able to take hold? (1 mark)

Hint 1: The descriptions should clearly explain what is occurring at each step.

Hint 2: Search through this list to find a description matching the process occurring at that point.

Hint 3: Here you are being asked to recall features of the difference between sexual and asexual reproduction.

Hint 4: You will need to study the steps in the diagram before answering the question.

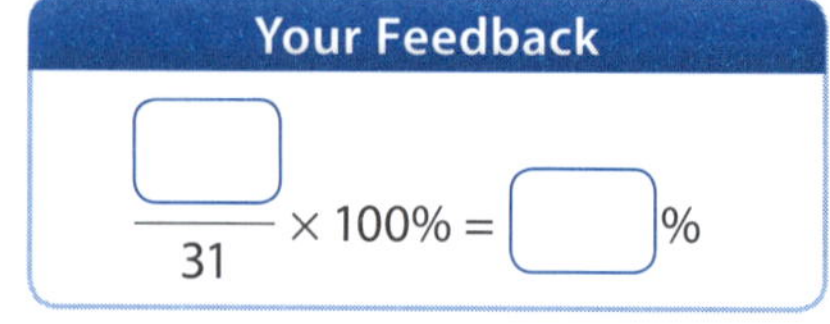

MATTER AND CHANGES OF STATE

Elements and compounds

QUICK REVISION

1 Matter is the general name given to the materials that make up the universe. Several thousand years ago, early Greek philosophers believed that all matter was made of very small __________ which they called atoms. This idea was revived by John Dalton in the early 19th century. With new technologies available to scientists, we can readily show the __________ nature of matter. Atoms have the ability to join together to form __________ particles called molecules. The __________ of matter in a body is measured by its mass. The unit of mass is the __________ or gram. Matter also occupies space. This is measured by the __________ of the material. The units of volume are litres, millilitres, cubic metres or cubic centimetres.

Matter can be converted into different physical __________. Water is a liquid at room temperature. If the water is __________ to 0 °C or below, it turns into ice which is a solid. If water is heated, the water begins to __________ and become a vapour. Water vapour is the gaseous form of water. Solids, liquids and gases are called the three states of matter. In solids the atoms or molecules are __________ closely together. The particles in solids are located in fixed positions. The particles __________ and the hotter the solid, the more the particles vibrate. Solids have a fixed shape but this shape can be __________ by hammering. The photo below shows a blacksmith hammering a hot sample of steel to create a horseshoe.

The volume of a solid is also fixed. In liquids the particles are still close together but the particles can __________ and slide around each other. Liquids have a __________ volume at a specific temperature but their shape depends on the vessel they are in. In gases the particles are very __________ apart and they move with great speed.

2 Atoms of different elements have different masses. The lightest elements are __________ and helium. Atoms of mercury and lead are very heavy. The volume of atoms also varies from one element to another. The __________ of a material depends on both the mass and the volume of the material. Density is defined as the __________ per unit volume and is measured in the units

(cont.)

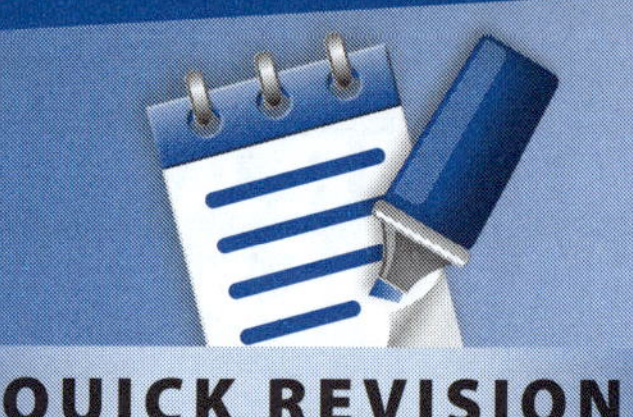

of g/cm³ or g/mL. Gases have quite __________ densities because they occupy larger volumes than liquids or solids. Solids are generally much __________ dense than liquids or gases. Iron, for example, has a density of 7.9 g/cm³ whereas water has a density of 1.0 g/cm³. This explains why iron objects __________ when placed in water.

Gases have the greatest ability to __________ from one place to another. The particles of gas move about at high speed and they can __________ with each other or solid materials. If a sample of gas is released into one corner of a closed room, it will spread out to __________ the room. This process is called __________. Diffusion can also be demonstrated using liquids of a different colour. If a red-dyed sample of water is injected at the base of a measuring cylinder of normal water, the red colour slowly __________ throughout the non-dyed water until the mixture has the same colour throughout. Solids do not diffuse into other solids at room temperature as the particles are __________ free to move about. The following diagram shows that an egg sinks in fresh water but floats in salt water (brine). Salt water is more dense than fresh water. The density of the egg is greater than the density of fresh water but less than that of salt water.

fresh water

brine

Gases can undergo compression and expansion. If a sample of gas is placed in a closed syringe and the plunger pushed in, the gas becomes __________ into a smaller volume. This happens because the gas particles have a lot of __________ space between them. During compression the particles move __________ to each other. The opposite happens when the plunger is withdrawn. Unlike gases, solids and liquids __________ be compressed. When all states of matter are heated, they do expand as the particles move apart.

3 The melting point of a material refers to the __________ at which a solid will turn into a liquid when heated. The freezing point is the same as the melting point but refers to the __________ process of turning a liquid back to a solid.

The boiling point of a material refers to the temperature at which its __________ will be turned into a vapour or gas. A liquid below its boiling point will still __________ to form a vapour. The higher the temperature, the more __________ is the evaporation. The opposite of evaporation is condensation. In this process a vapour __________ and turns back into a liquid. Thus water vapour turns back into drops of liquid water as it makes contact with a cold surface.

Answers **1** particles; atomic; larger; amount; kilogram; volume; states; cooled; evaporate; packed; vibrate; changed; slip; fixed; far **2** hydrogen; density; mass; low; more; sink; spread; collide; fill; diffusion; diffuses; not; compressed; empty; closer; cannot **3** temperature; reverse; liquid; evaporate; rapid; cools

MATTER AND CHANGES OF STATE

Elements and compounds

REVISION SUMMARIES

1 The universe, and all the materials in it, is made up of matter. **Matter** is composed of extremely small particles called **atoms**. Various atoms join together to form larger particles. There are two important characteristics of matter.

- Matter occupies space. All materials have a **volume** which is measured in units such as litres (L) and millilitres (mL) for liquids and gases and cubic metres (m^3) or cubic centimetres (cm^3) for solids.
- Matter has **mass**. This is the amount of matter present and is measured in units such as kilograms (kg), grams (g) and milligrams (mg).

Matter can exist in three different states called solids, liquids and gases. These groups are based on the arrangement and motion of their atoms and are called the states of matter.

- In **solids**, atoms are closely packed together. Solids have a fixed shape and volume and the particles in the solid remain in fixed positions but they do vibrate.
- The particles in **liquids** are very close to each other but are able to slide over one another as they move from place to place. Liquid particles take the shape of the lowest part of the container they are in. But liquids do have a fixed volume.
- The particles of **gases** are spread apart. Gases have neither fixed volume nor shape. The gaseous particles move quickly and fill the entire container they are in.

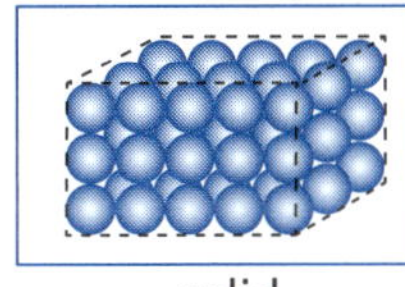

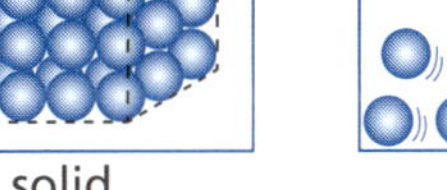

solid

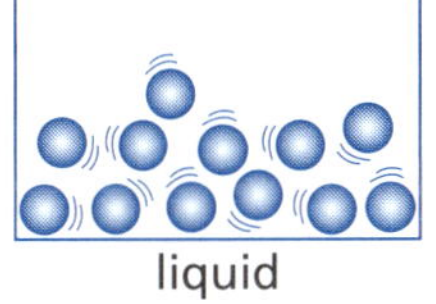

liquid

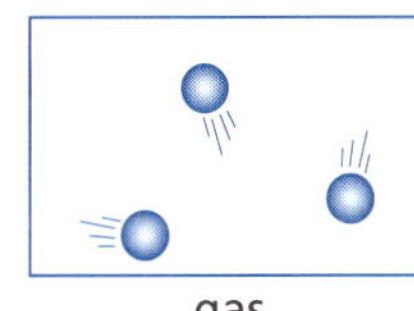

gas

2 The particle model of matter helps us to understand the properties of solids, liquids and gases.

- **Density.** The density (D) of a material is defined as the mass (m) per unit volume (V).

$$D = \frac{m}{V}$$

Solids generally have the greatest density as the particles are closely packed together. Solids vary considerably in their density as some atoms are much heavier than others. Gold atoms are about eight times heavier than sodium atoms. These atoms have different volumes and pack in different ways in their crystal. Gold has a density of 19.3 g/cm^3 whereas sodium's density is 0.97 g/cm^3. This is slightly less than liquid water, which has a density of 1 g/cm^3. Solid ice, however, is less dense than liquid water as the particles in the ice are not packed closely together. Ice therefore floats in water. The photo shows ice blocks floating in water. Ice and water are the same chemical substance yet ice is less dense than liquid water due to the way the water molecules pack together in the ice crystal. Gases have the lowest density as their particles are spread apart from each other.

(cont.)

- **Diffusion.** Diffusion is the process in which the particles of matter spread out from a zone of high **concentration** to a zone of low concentration. Gases diffuse readily into other gases as their particles are far apart. This is demonstrated by the movement of gaseous perfume molecules through the air when the lid of the perfume bottle is removed. The particles of liquids can also diffuse from one region to another but the rate of diffusion is slower than in gases. Solids demonstrate no diffusion as the particles do not move from one place to another.
- **Compression.** Gases can be readily compressed into a smaller volume. This process reduces the free space between the particles. Liquids and solids are **incompressible** as there is no free space between the particles.
- **Expansion.** Gases will expand if we increase the volume of a container they are in whereas this does not occur with liquids and solids. All matter will expand if heated as the particles become more energetic. Gases show the greatest degree of expansion when heated. The expansion of mercury or alcohol has been used in devices such as thermometers.

3 Matter can change between the three states. If a block of ice at 0 °C is left on a tray in sunlight, it quickly melts (or fuses) to form liquid water. Eventually the liquid water evaporates. Water molecules enter the air as water vapour. You can speed up the **evaporation** (or vaporisation) process by heating the liquid water in an electric kettle. When the liquid water reaches 100 °C, it boils. When a liquid boils, bubbles of vapour rise to the surface. When water is cooled, changes in state happen in the reverse direction. Water vapour cools and turns into liquid water. Liquid water cools and forms ice. **Condensation** (or liquefaction) of water vapour, for example, forms on a cold bathroom mirror while you are showering. Solidification (or freezing) of liquid water to form ice, for example, happens when you make ice blocks in a freezer.

Checklist

Can you: ✓

1. *Distinguish between solids, liquids and gases in terms of the arrangement of particles?* ☐
2. *Use the particle theory of matter to describe the various properties of solids, liquids and gases?* ☐
3. *Describe the energy processes involved when solids, liquids and gases are converted from one state to another?* ☐

MATTER AND CHANGES OF STATE

Elements and compounds

REVISION TEST

1 **a** Use the following diagram to measure the mass of material in three vessels of the same volume. (3 marks)

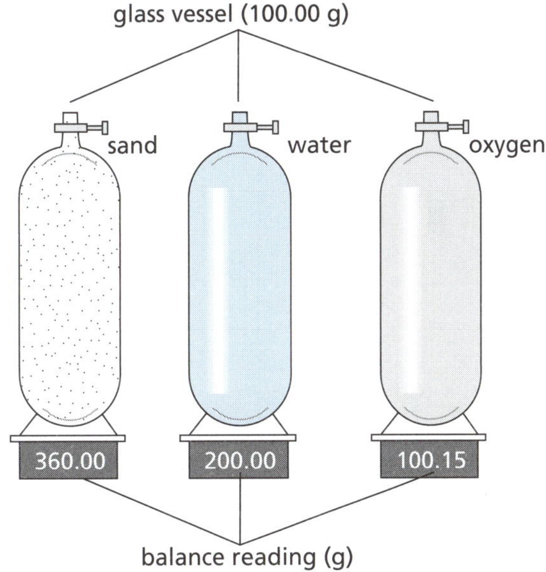

b Which material has the greatest density and which has the least? *Hint 1* (2 mark)

2 A student placed hundreds of small steel ball bearings in a plastic container. She then agitated the container from side to side as shown in the diagram below. What state of matter is she modelling? (1 mark)

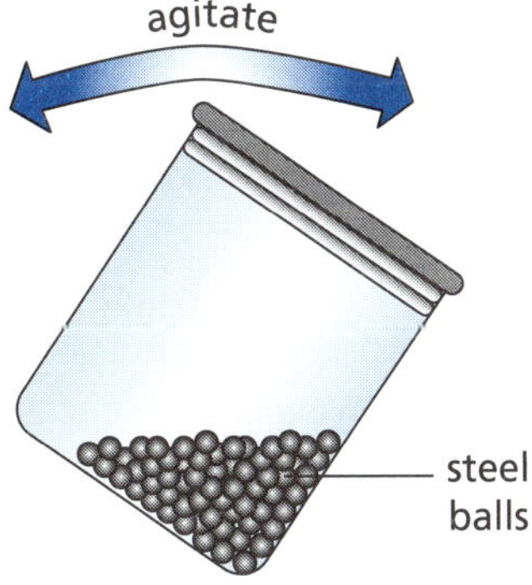

3 The following diagram shows an experiment in which a student places water in a syringe and then pushes against the plunger.

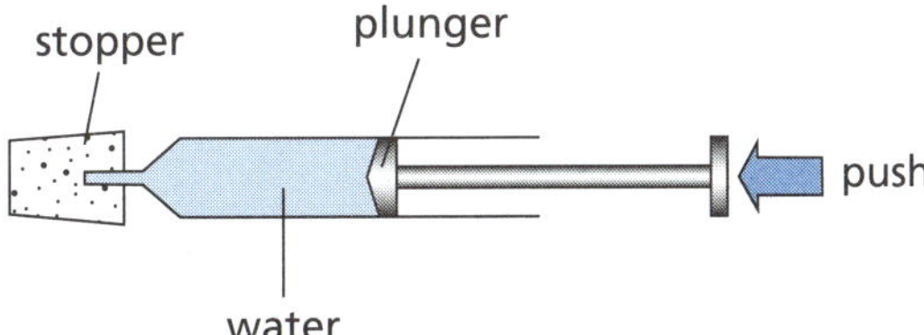

a Describe what will happen and explain the student's observations. (2 marks)

b The experiment is repeated with the water replaced by air. Explain what will happen in this experiment. (2 marks)

(cont.)

4 The atmosphere on the surface of Mars is much thinner than on Earth. What can you say about the number of particles in a given volume of atmosphere on Mars compared to the same volume of atmosphere here on Earth? *Hint 2* (1 mark)

5 The following diagram shows a sample of yellow-green chlorine gas being injected into the base of a closed vessel of air. After some time the yellow-green gas fills the container.

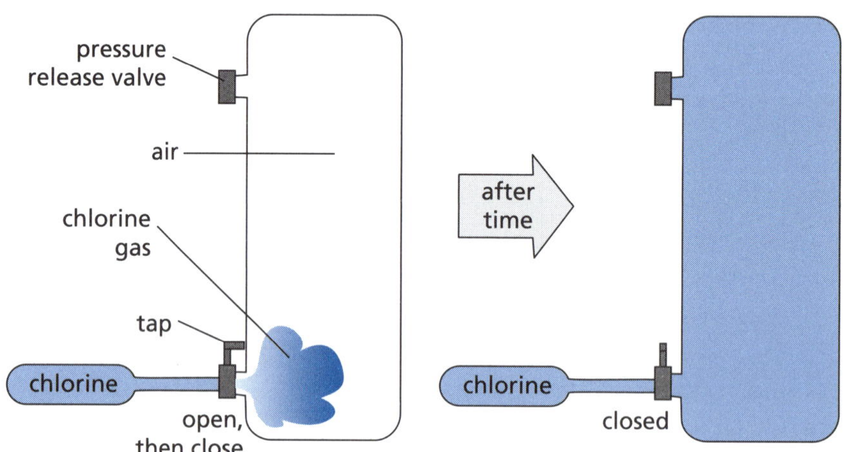

a Name the process that allows the chlorine to spread throughout the container of air. (1 mark)
b Explain how this process occurs. (1 mark)

6 Calculate the density of silver metal from the following data. (2 marks)
mass of silver = 52.5 g
volume of silver = 5.0 cm^3

7 The melting point of mercury is –39 °C and its boiling point is +357 °C.
a In what state is mercury at 25 °C? *Hint 3* (1 mark)
b Some mercury vapour is cooled from 400 °C to 300 °C.
i What would you expect to see happen during the cooling? (1 mark)
ii What name is given to the process you observe in this experiment? (1 mark)
iii Name the process in which liquid mercury is converted into solid mercury. (1 mark)

Hint 1: What is the formula for density? Remember that the volumes are all the same.
Hint 2: Thin air means that there are fewer particles.
Hint 3: A substance is a solid below its melting point and a vapour above its boiling point.

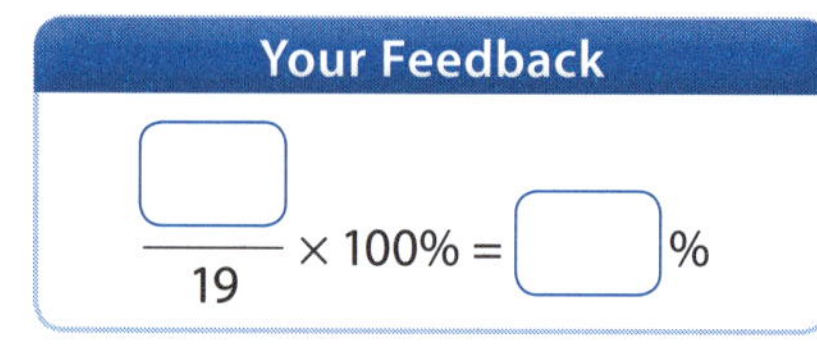

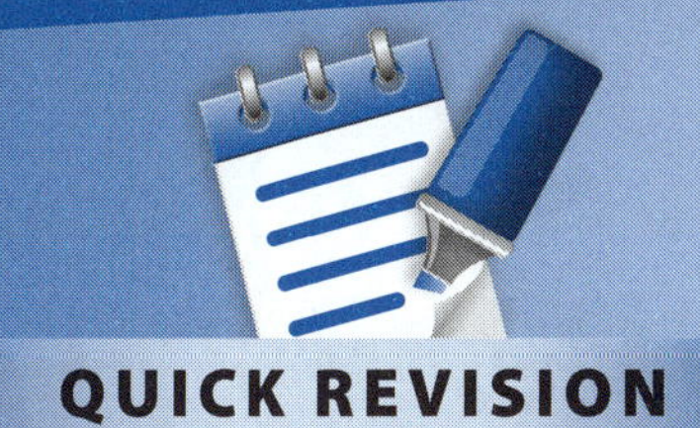

ELEMENTS

Elements and compounds

QUICK REVISION

1 The water from our tap is classified as a __________ as it contains water, fluoride, chlorine and air. Most materials in the world are mixtures. These mixtures can be separated using __________ separation techniques. Pure substances cannot be separated by physical separation techniques such as filtration, distillation and chromatography. The simplest types of __________ substances are called elements. Elements cannot be turned into anything __________ by chemical processes. Every element in the universe is unique. The most abundant and lightest element in the universe is __________. Nitrogen is an element that represents about 78% of our atmosphere. Nitrogen atoms have a mass that is about 14 times greater than hydrogen atoms. The weight of an atom is mainly due to the mass of its nucleus. The nucleus of an atom contains two types of particles called __________ and neutrons. Protons have a positive charge and neutrons have no charge. These nuclear particles are much heavier than the lightweight __________ that are negatively charged and which orbit the nucleus. Iodine, for example, is an atom with 53 protons, 74 neutrons and 53 electrons. Protons and neutrons are each about 1840 times __________ than an electron.

2 Each element has a unique symbol and a unique __________ number (Z). The atomic number of an atom is equal to the number of __________ in its nucleus. Hydrogen has an atomic number of 1 and helium has an atomic number of 2. Uranium, which is a very heavy element, has an atomic number of 92. The symbol for an element is either __________ or two letters, with the second letter being lower-case. Thus C is the symbol for carbon, __________ is the symbol for calcium and Cl is the symbol for chlorine. Some symbols are based on __________ names such as Pb for *plumbum* and Fe for *ferrum*. Some elements are named to honour scientists. For example, Cm is for curium and Rf for rutherfordium. Not all elements are natural. In recent years some __________ elements have been created in particle accelerators. Such elements are __________ and break down into lower mass elements. Californium (Cf) is a synthetic element. The elements are arranged in a classification scheme called the __________ table. A copy of this table is on the inside cover of this book. This table arranges elements in order of __________ atomic number. The vertical columns are called __________ and they correspond to families of related elements. The elements in the first column or group all have __________ electron in their outermost orbit around the nucleus.

3 The periodic table arranges the elements in order of increasing atomic number (Z). Most of the elements are classified as __________. Metals have some common characteristics. Metals are lustrous and malleable. Malleability refers to the ability of a solid to be __________ or hammered into sheets. Aluminium foil is an example of this property. Metals are also __________ which means that the material can be drawn out or stretched into thin wires. Copper wire is such an example. Metals are also good __________ of electricity and heat. Saucepans are made of metals as they are good heat conductors. Non-metals have the __________ properties to metals. They are located on the right-hand side of the periodic table. Oxygen, sulfur and phosphorus are examples of non-metals. The __________ gases such as helium and argon are also non-metals. Non-metals are dull in appearance and they are __________ if hammered. Non-metals are electrical and __________ insulators. Semi-metals have properties in between those of __________ and non-metals. Silicon and germanium are examples of semi-metals. Silicon is the basis of silicon chips that are used in modern computers.

Answers **1** mixture; physical; pure; simpler; hydrogen; protons; electrons; heavier **2** atomic; protons; one; Ca; Latin; synthetic; unstable; periodic; increasing; groups; one **3** metals; rolled; ductile; conductors; opposite; noble; brittle; heat; metals

1 In the study of chemistry, the term ***pure substance*** refers to materials that have a constant composition. Most materials in the world are not pure as they contain variable amounts of other substances. Natural water is not pure as it has some air dissolved in it. Some materials that can be found in the pure state are gold, sulfur and platinum. Gold and platinum are very unreactive and consequently rarely combine with other substances.

The simplest of the pure substances are called **elements**. Elements cannot be converted into anything simpler by physical or chemical means. The atoms of one element are all identical to each other. Different elements have atoms of different mass. Hydrogen has the lightest atoms of all the elements. Gold atoms are much heavier with an atomic weight 197 times greater than hydrogen. Oxygen atoms have an **atomic weight** which is 16 times greater than hydrogen. These atoms have different atomic weights because they have different numbers of subatomic particles. Atoms are composed of different numbers of protons, neutrons and **electrons. Protons** and **neutrons** have similar mass and are much heavier than electrons. Protons and neutrons are found in the atom's nucleus which represents most of the atom's mass.

Examples:
composition of gold atoms: 79 protons + 118 neutrons + 79 electrons
composition of oxygen atoms: 8 protons + 8 neutrons + 8 electrons

most of the mass of the oxygen atom is in the nucleus

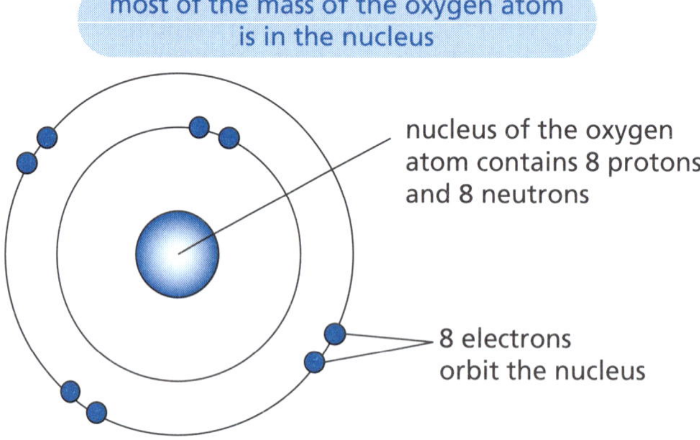

2 The materials of the universe are composed of different elements. **Hydrogen** is the most abundant element of the universe. Each element is given a unique symbol. The symbol for an element is made up of one or two letters (the first letter is always a capital). Thus H is the symbol for hydrogen but He is the symbol for helium. Some elements have symbols related to their ancient name or name in a different language. For example, Cu is the symbol for copper as its Latin name is *cuprum*.

Elements are classified into **groups** or families. Members of a family of elements have similar chemical properties and a gradation in their physical properties. The arrangement of electrons in their atoms also shows a regular pattern. The periodic table (see the inside cover of this book) shows these families of elements as vertical columns called groups. Thus, O, S, Se, Te and Po are a family of elements (find this group on the periodic table). Each of these elements has 6 electrons in the outermost atomic orbit, and this allows them to have similar chemical properties. Some of the elements shown in this table are synthetic and quite unstable. They have been created by physicists using giant particle accelerators. Element 99 (Einsteinium, Es), for example, is a synthetic element named in honour of Albert Einstein. The last group of the periodic table contains a family of very unreactive gases called the noble gases. Helium is the most unreactive member of this family.

The elements of the **periodic table** are arranged in order of the number of protons in their nucleus. This number is called the **atomic number** (Z). The first element hydrogen (H) has an atomic number of 1.

3 Elements can be further classified into groups called metals, semi-metals and non-metals. The following diagram shows a plan map of the locations of metals, semi-metals and non-metals in the periodic table. Most elements are metals. The semi-metals have properties that are intermediate between metals and non-metals.

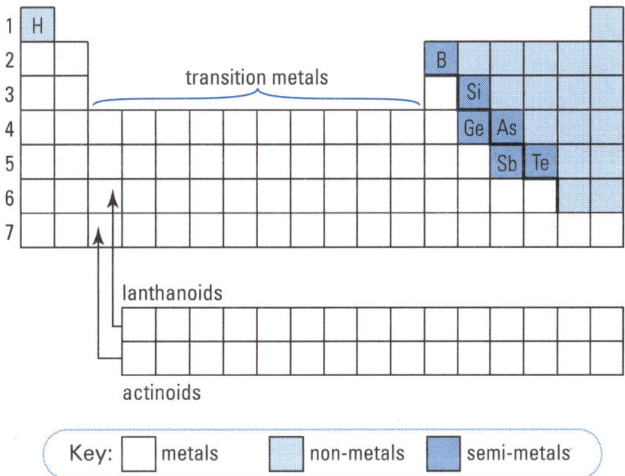

Metals can be distinguished from **non-metals** by their physical and chemical properties. Metals are malleable. This means they can be hammered or rolled into thin sheets. They are also ductile which means they can be drawn into wires. Non-metals are brittle which means they shatter on hammering. The following table shows properties of metals and non-metals.

Properties of metals	Properties of non-metals
shiny surface	dull surface
ductile	brittle
malleable	non-malleable
good conductor of heat	poor conductor of heat
good conductor of electricity	poor conductor of electricity, except carbon (graphite)

Silicon is an example of a semi-metal. It resembles a metal as its surface is grey and lustrous. It is a good heat conductor but it is only a semi-conductor of electricity.

Checklist
Can you:

1 *Explain why elements are pure substances?* ☐
2 *Describe the families of elements in the periodic table?* ☐
3 *Distinguish between the properties of metals and non-metals?* ☐

REVISION TEST

1 The following apparatus is used to test the electrical conductivity of different elements.

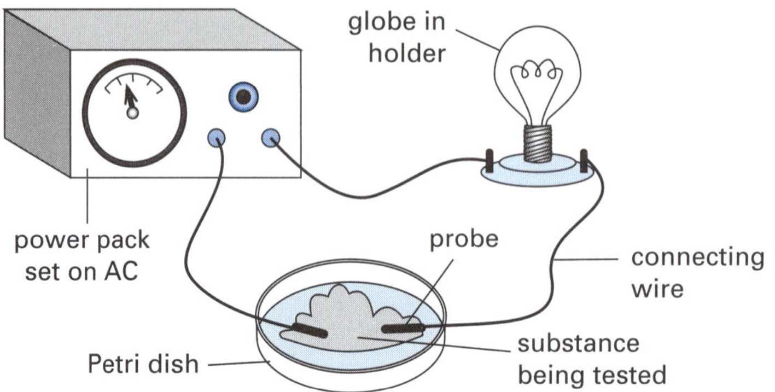

a What is the purpose of the light globe? (1 mark)

b Name a material that could be used as the probes in this experiment. *Hint 1* (1 mark)

c The materials tested were sulfur, zinc, iodine, iron, nickel, phosphorus and calcium. Identify which of these materials show good electrical conductivity. (4 marks)

2 Darmstadtium is a synthetic element that is made using a particle accelerator. Charged nickel atoms were accelerated and they collided with lead atoms. This collision yielded an atom of darmstadtium as well as a neutron, as shown in the diagram below. The numbers associated with each atom refer to their relative atomic masses (unit = atomic mass unit, u).

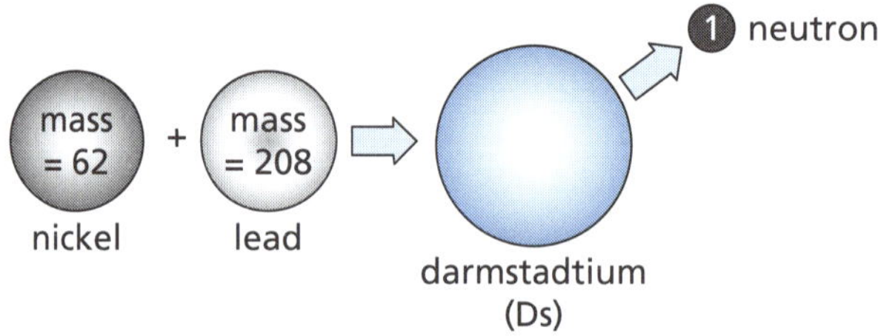

a Use the periodic table to write the symbols for nickel and lead. (2 marks)

b Use the information in the diagram above to calculate the mass of a darmstadtium atom assuming the total mass of reactants is the same as the total mass of products. (1 mark)

c Nickel has 28 protons in its nucleus and lead has 82 protons. How many protons are there in the nucleus of the darmstadtium atom? (1 mark)

d Determine the atomic number of darmstadtium. *Hint 2* (1 mark)

3 Match the elements in the left column with their chemical symbol in the right column. *Hint 3* (8 marks)

Element	Symbol
A iron	1 Zn
B copper	2 C
C carbon	3 Cu
D helium	4 Fe
E selenium	5 Hg
F mercury	6 Cl
G zinc	7 He
H chlorine	8 Se

4 Use the information in the table below to complete the flow chart in the figure that follows. In some cases the distinguishing characteristic has been left out; in others the name of the element has been omitted. (12 marks)

Element	Discovered by	Features	Density (g/m³)	Melting point (°C)	Boiling point (°C)
cobalt	G. Brandt, Sweden 1735	very hard, pinkish-white metal, tough, extremely magnetic; used in cancer treatment in radioactive form	8.9	1490	2900
tungsten	J. and F. de Elhuyar (brothers), Spain 1783	very hard, silvery-white metal, resistant to chemicals; used in tungsten filament lamps, alloyed in steel, cutting tools tipped with tungsten carbide	19.3	3410	5930
sodium	H. Davy, Britain 1807	soft, silvery-white metal, extremely active and has to be stored under oil; used in vapour lamps	1.0	98	892
zinc	A. von Svab (1742) and A. Marggraf (1746), Sweden	bluish-white metal, soft, reacts with air; used as outer casing for dry cells, toys, coating for steel	7.1	419	906
iron	unknown, smelted first in 2700 BC	grey-white metal, hard and tough, combines easily with oxygen; used extensively in industry	7.9	1540	3000
magnesium	H. Davy, Britain	silvery metal, surface reacts with air, burns with a bright flame; used in fireworks, flares, important component of plant pigment chlorophyll	1.7	680	1110

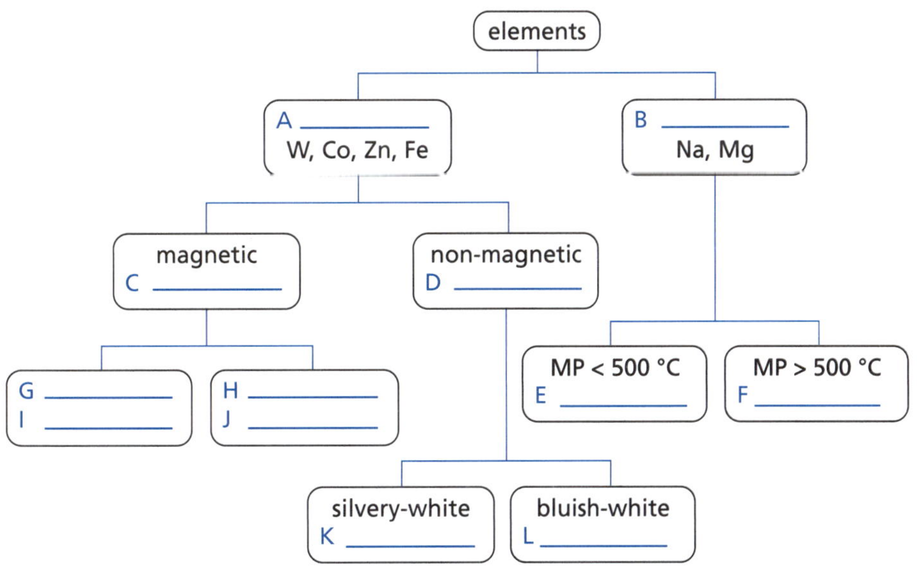

Hint 1: The probes must be able to conduct electricity.
Hint 2: What is the connection between atomic number and the number of protons in the nucleus?
Hint 3: Some elements have symbols related to their Latin names.

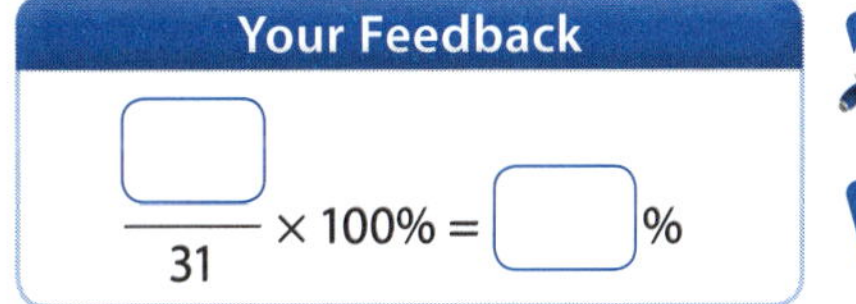

PAGE 189
PAGE 212

COMPOUNDS

Elements and compounds

QUICK REVISION

1 Just as some of the 26 letters in the English __________ can combine together to produce the many thousands of different __________ in this book, atoms of chemical __________ can combine together to form many different __________. A chemical compound is a pure substance consisting of two or more elements __________ joined together. These can be separated into simpler substances by various chemical __________. The elements cannot be separated in a compound by __________ separation methods.

The physical and chemical properties of a compound are __________ from the elements that make it up. That is, when elements join to become __________, they lose their individual characteristics. For example, sodium metal is very reactive and chlorine is a poisonous __________. But when sodium and chlorine combine, they form a non-reactive substance called sodium chloride (__________ salt, NaCl). This compound, frequently used in cooking, has none of the __________ of the original elements. The new compound, sodium chloride, is not as reactive as are the original __________.

2 Elements join together in __________ by forming chemical bonds that link the atoms. This can be done in several ways. One way is for atoms to share their outermost electrons forming a stable relationship. Such bonds are called __________ bonds. Non-metals join together like this to form __________ compounds. For example, in water the hydrogen atom shares its electron with one from the __________ atom.

Another way is for one element to __________ its outermost electrons to another element. This makes both elements electrically __________ and these atoms are now called __________. The ionic compounds formed are linked together by the __________ charges they produce. This is typical for a __________ joining with a non-metal. For example, in sodium chloride the sodium atom can __________ one electron to a chlorine atom. As the sodium atom now has one __________ electron, it is positively charged. As the chlorine atom now has one __________ electron, it is negatively charged. These opposite charges __________ each other.

The following diagram shows a model of sodium chloride crystals in which there are equal numbers of sodium ions and chloride ions.

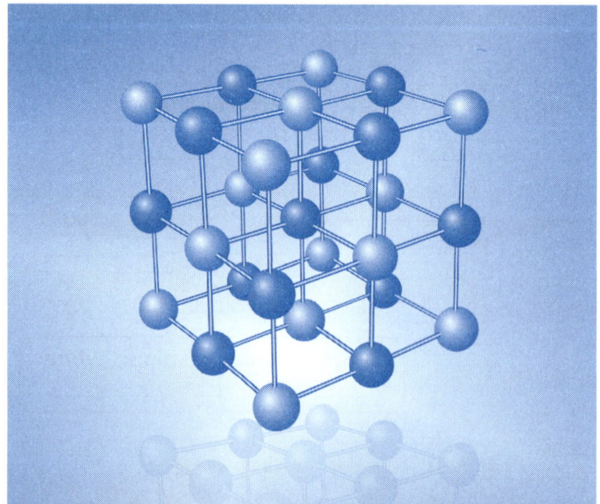

3 Compounds exist all around us. For example, we cook with salts such as table __________ (sodium chloride) and baking __________ (sodium bicarbonate), we have carbon __________

in our soft drinks to make them fizz and we put __________ (sucrose) in our tea. We soothe an upset __________ by using antacids containing magnesium hydroxide and aluminium hydroxide when there is too much __________ acid, a strong acid used to break down food, in the stomach. We clean with __________ that often contains sodium hypochlorite or sodium perborate, wash with washing soda (sodium __________), put sulfuric acid in the __________ of our cars and put a mixture of carbon and hydrogen __________ in the petrol tanks. We make statues and bench tops from __________ (calcium carbonate). Even nail polish __________ is a simple chemical compound called acetone.

The following photo shows a student model of carbon dioxide which consists of one atom of carbon joined to two atoms of oxygen.

4 Plastics are very common in the 21st century, although they were little known over a century ago. Today plastics are __________, versatile and form part of many materials around us. A plastic material is any of a wide range of synthetic solids that can be __________. Plastics are typically long-chain __________ (polymers) of carbon and hydrogen but often containing other elements. They are most commonly derived from __________, while some are partially natural.

Worldwide, plastic production is now over 80 million __________ a year. Most plastic is not __________ and will persist in the environment for hundreds of years. For example, plastic films last for up to 30 years while __________ bottles can last indefinitely. Each year discarded plastics kill __________ of sea birds, sea mammals and fish as many get entangled in plastic six-pack rings, plastic strapping and nylon ropes. However, most plastics can be __________ so it is important we recycle them whenever we can. For those that can't, we should __________ of them properly and not allow them to get into waterways.

Answers 1 alphabet; words; elements; compounds; chemically; reactions; physical; different; compounds; gas; table; traits (characteristics, features); elements 2 compounds; covalent; molecular; oxygen; donate (give up); charged; ions; opposite; metal; donate; less; more; attract 3 salt; soda; dioxide; sugar; stomach; hydrochloric; bleach; carbonate; batteries; compounds; marble; remover 4 cheap; moulded; molecules; petrochemicals (oil industry); tonnes; biodegradable; plastic; millions; recycled; dispose

COMPOUNDS

REVISION SUMMARIES

1 As there are only around 90 naturally occurring elements, but millions of compounds, most substances are therefore not found as independent atoms of elements. Atoms can form groups, called **molecules**, in compounds. Compounds contain two or more different elements in a chemically combined form. The simplest form of a compound is a molecule. A molecule contains two or more atoms chemically bonded together. A molecule is the smallest particle of a pure chemical substance that still retains its composition and chemical properties. For example, a molecule of water always contains two hydrogen atoms and one oxygen atom (written as H_2O). Another compound, also containing the elements hydrogen and oxygen, is hydrogen peroxide. Its molecule can be written as H_2O_2.

H H O

molecule of water (H_2O)

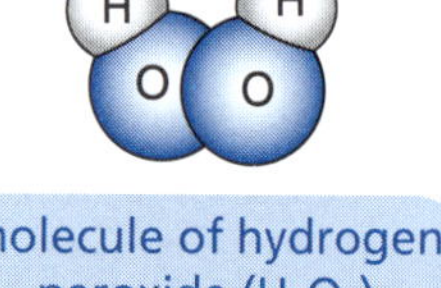

molecule of hydrogen peroxide (H_2O_2)

The composition of compounds is indicated by a **chemical formula** using symbols for the elements. The number of atoms of each element in a formula is given by a subscript. If no subscript is given, then the number of atoms is one. For example, a molecule of the gas carbon dioxide can be written as CO_2. Each molecule consists of three atoms, two of which are oxygen and one of which is carbon.

The ratio of elements always remains the same for any particular compound. For water, the ratio of H:O is 2:1). This is true for all water molecules. In fact, all compounds have the same ratio of the elements that make them up, and are pure substances. Some molecules consist of only one element; for example, oxygen gas (O_2), nitrogen gas (N_2) and hydrogen gas (H_2). Ozone (O_3) is a form of oxygen consisting of three oxygen atoms bonded together.

2 Compounds are very common in our immediate environment. The following table gives some compounds and their common uses.

Chemical compound	A common use
sodium fluoride	present in toothpaste
acetone	nail polish remover
ethanol	alcohol present in alcoholic drinks
phosphoric acid	found in many soft drinks
butane	the lighter fuel in cigarette lighters
methane	main compound found in cooking gas piped to homes
sodium lauryl sulfate	present in soaps
sucrose	common sugar present in foods
sodium nitrate	found in garden fertiliser
octane	a main component found in petrol
sodium hypochlorite	found in bleach
acetic acid	vinegar used in cooking
aspirin	the active ingredient is acetylsalicylic acid
sodium chloride	table salt found in foods
methyl salicylate	oil of wintergreen found in muscle ache creams

3 Chemical compounds are formed by joining two or more atoms. In **chemical bonds**, atoms can either share or transfer some of their electrons. A stable compound occurs when the total

energy of the combination is lower than the separate atoms. In this bound state there is a net attractive force between the atoms keeping them together. The following are two important examples of chemical bonds.

- **Covalent bond.** This is where one or more pairs of electrons are shared by two atoms. Such bonds lead to stable molecules. This often occurs between non-metal elements. The combination produces molecular compounds. Molecular compounds have low melting points and do not conduct electric currents.
- **Ionic bond.** This is where one or more electrons are removed from one atom and attached to another atom. This produces positive and negative ions (charged atoms) which attract each other. The combination produces ionic compounds. Ionic compounds are generally solids with high melting points and conduct electrical currents when melted or dissolved in water.

How strong the bond is depends on several factors, such as the charges present on each ion and the distance between them. Small, highly charged ions produce strong bonds while large, minimally charged ions produce weaker bonds. The bond strength can be measured by how much energy is needed to tear apart the atoms or ions from each other.

4 Celluloid was one of the first plastics, and was invented in 1868. Through experimentation, the new celluloid could be moulded with heat and pressure into a durable shape. Celluloid became famous as the first flexible photographic film used for still photography and motion pictures. Today **plastics** are found in fine wrapping film, plastic bags and containers and stiffer plastic boxes, lids and other containers. Your calculator has its electronics inside a stiff plastic casing designed to protect the delicate components.

```
H       H           H  H                           H  H  H  Cl H  H  H  Cl H  H  H  Cl
 \     /            |  |    polymerisation         |  |  |  |  |  |  |  |  |  |  |  |
  C = C    --->    -C--C-   -------------->   ... -C--C--C--C--C--C--C--C--C--C--C--C- ...
 /     \            |  |                           |  |  |  |  |  |  |  |  |  |  |  |
H       Cl          H  Cl                          H  Cl H  H  H  Cl H  H  H  Cl H  H
```

simple molecule is the monomer — one covalent bond breaks and reaction begins — a long chain polymer molecule forms

Plastics are made up of long chain molecules called **polymers**. Many organic polymers are plastics. Most of these polymers are based on chains of carbon atoms alone or with oxygen, sulfur or nitrogen as well. Chemists are able to customise the properties of a plastic by hanging different molecular groups or atoms from the carbon backbone. The structure of these side chains influences the properties of the polymer.

Polystyrene is one type of plastic formed by joining together many smaller styrene molecules. It can form hard impact-resistant plastics for furniture, cabinets, cups and eating utensils. When polystyrene is heated and air blown through the mixture, it forms styrofoam. Styrofoam is lightweight, mouldable, an excellent insulator and is used in foam cups to hold hot drinks.

Checklist

Can you:

1. *Explain why compounds are pure substances?* ☐
2. *Identify types of compounds and their structure, including molecular compounds and ionic compounds?* ☐
3. *Identify some uses of common compounds?* ☐
4. *Explain that plastics are large molecules made from small molecules?* ☐

COMPOUNDS

Elements and compounds

REVISION TEST

1 Complete the following statements.

a A substance that cannot be broken down into a simpler form by ordinary physical means is called a __________ substance. (1 mark)

b When an atom gains or loses electrons it is then called an __________. (1 mark)

c A molecule is the smallest complete unit of a __________. (1 mark)

d The written notation used to represent the number of atoms of each element in a compound is called its chemical __________. (1 mark)

e A chemical substance that has a molecular structure built up from a large number of similar units bonded together is called a __________. (1 mark)

2 True or false?

a The proportions of the elements in a compound are always fixed. (1 mark)

b The elements of a compound retain their individual properties. (1 mark)

c Compounds are made up of two or more different elements. *Hint 1* (1 mark)

d It takes a large amount of energy to separate the elements in a compound. (1 mark)

3 The law of constant composition states that the ratio by mass of the elements in a chemical compound is always the same, regardless of how the compound is made or from where it came. Explain this statement. (2 marks)

4 Lead sulfate ($PbSO_4$) is a white crystal or powder found in the battery of a car.

a Name the elements present in lead sulfate. *Hint 2* (3 marks)

b What is the ratio Pb:S:O? (1 mark)

5 Table sugar (sucrose) has the formula $C_{12}H_{22}O_{11}$.

a Is this substance an element, compound or mixture? (1 mark)

b Name the different atoms present in the sugar molecule. (3 marks)

c How many atoms in one molecule of sugar? (1 mark)

d Another compound has the formula $C_6H_{12}O_6$. Is this sucrose as well? Explain. (2 marks)

e What is the ratio C:H:O in $C_6H_{12}O_6$? *Hint 3* (1 mark)

6

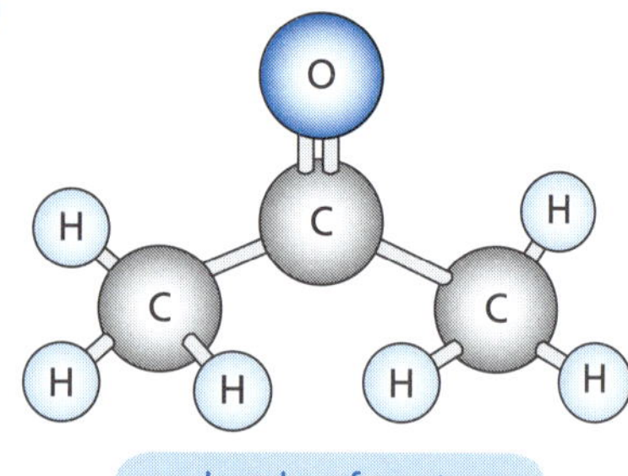

molecule of acetone

The diagram shows a molecule of acetone, a colourless and flammable liquid commonly used as nail polish remover.

a What elements are present in acetone? (1 mark)

b The chemical formula for acetone can be written as $C_pH_qO_r$. What are the values of p, q and r? *Hint 4* (3 marks)

c What do the links between the atoms represent? (1 mark)

d Are the atoms in this molecule joined together by covalent or ionic bonds? How do you know? (2 marks)

7 a Petrol used in cars is a mixture of several compounds. All are colourless. Yet premium unleaded petrol is yellow, while regular unleaded is purple. Explain. *Hint 5* (1 mark)

b Give two features these colouring substances must have. (2 marks)

8 The following diagram shows how covalent and ionic bonds form.

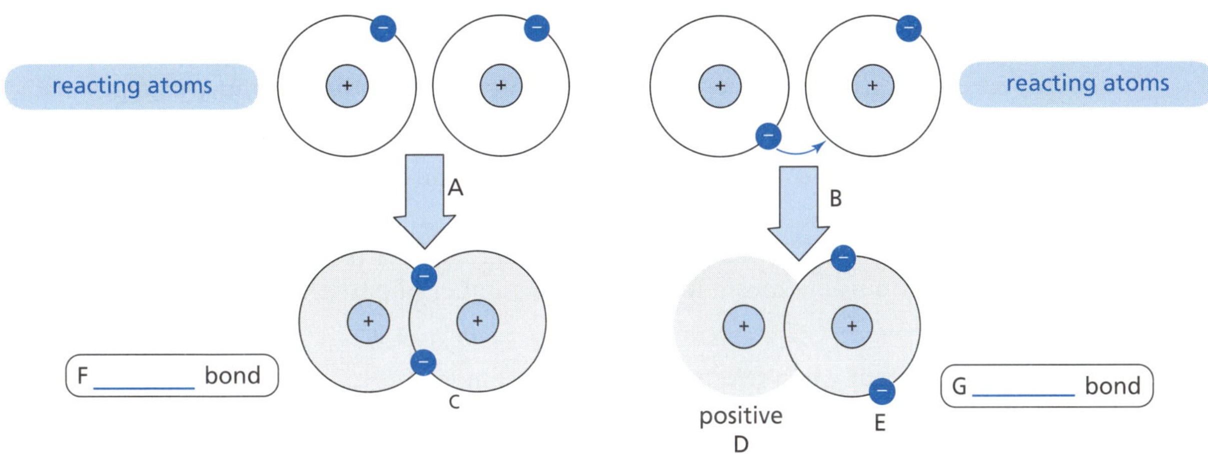

a What is a chemical bond? (1 mark)

b Replace the letters A to G on the diagram with words or phrases from the following list: ion; covalent; two atoms share their electrons; negative ion; a molecule forms; one atom transfers its electron to another atom; ionic. (7 marks)

9

```
    H   H       H   H
    |   |       |   |
H — C — C — O — C — C — H
    |   |       |   |
    H   H       H   H
```

ethoxyethane (ether)

```
    H   H   H   H
    |   |   |   |
H — C — C — C — C — O — H
    |   |   |   |
    H   H   H   H
```

butanol

The diagram shows the molecules of two different compounds: ethoxyethane (commonly known as ether) and butanol. The '—' between atoms indicates a chemical bond linking two atoms.

a How many of each atom is present in a molecule of ethoxyethane? (3 marks)

b How many of each atom is present in a molecule of butanol? (3 marks)

c Besides the constituent elements being in the same proportion, what other feature distinguishes these two compounds from each other? (1 mark)

10 The following diagram shows smaller chemical units combining to form a larger unit.

a What is shown in this process? (1 mark)

b What type of chemical substances form by this process? (1 mark)

Hint 1: There is a subtle distinction here. Consider the word 'different'.

Hint 2: You may use the periodic table to find the names of the elements given.

Hint 3: Simplify the numbers in the ratio, just as you would the numbers in a fraction.

Hint 4: The letters p, q and r represent the number of atoms of each element in a molecule of the compound.

Hint 5: The colouring agent has been deliberately added to petrol. Why do you think this is done?

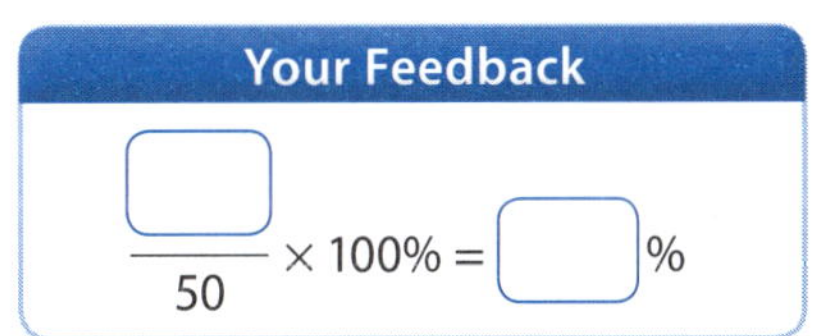

PAGES 189–190

PAGE 212

SIMPLE CHEMICAL REACTIONS

Elements and compounds

QUICK REVISION

1 When an ice cube ________, for example, it turns into liquid water. The solid ice and liquid water have the same ________ composition. The only difference is the form: one is solid, the other is ________. This is called a ________ change. In a chemical change, substances are altered into ________ substances. That is, the composition of the substance changes.

These chemical changes are also referred to as chemical ________. In chemical reactions different substances combine. New substances with new and different physical and ________ properties are formed.

Matter is never destroyed or ________ in chemical reactions. The particles of one substance are rearranged to form a new substance. The same number of particles that existed before the reaction exist ________ the reaction.

2 Synthesis reactions occur when two or more simple compounds ________ to form a more complex one. This can be written simply as A + B → AB.

- The ________ are shown on the left-hand side of the arrow. These are the chemical substance reacting together.
- The ________ (the substance or substances formed) are written on the right-hand side of the arrow.
- The arrow (→) in a chemical ________ stands for 'react to produce'.

For example, when hydrogen is burnt in air (in the presence of ________) water forms. This can be written as a word equation:

hydrogen + oxygen → water

Another example of a synthesis reaction is the formation of potassium chloride from potassium and chlorine gas. The word equation is:

potassium + ________ → potassium chloride

3 The reaction between a non-metal and a non-metal produces ________ compounds. This is where the non-metals ________ electrons in the compound formed. The reaction between ________ and oxygen forming water gives rise to a molecular compound such as water.

The reaction between a metal and a non-metal produces ________ compounds. This is where the metals ________ one or more electrons to the non-metal in the compound formed. The reaction between ________ and chlorine forming potassium chloride forms ionic compounds.

4 In a decomposition reaction a single compound is broken down into two or more ________ compounds or elements. That is, a more ________ substance is broken into simpler ones. In some ways this reaction is opposite to that of a ________ reaction. This can be written as AB → A + B.

For example, when electricity is passed through water it can be decomposed into its components:

water → hydrogen + ________

Answers **1** melts; chemical; liquid; physical; different; reactions; chemical; created (produced); after **2** combine; reactants; products; reaction; oxygen; chlorine **3** molecular; share; hydrogen; ionic; transfers; potassium **4** smaller; complex; synthesis; oxygen

SIMPLE CHEMICAL REACTIONS

Elements and compounds

REVISION SUMMARIES

1 In physical changes, no new substance is formed. **Physical changes** are usually easy to reverse. For example, after ice melts into liquid water, it can be refrozen into solid ice by lowering the temperature. Chemical changes involve the formation of new substances. These new substances have different physical and chemical properties. That is, the composition of a substance does change. Many **chemical changes** (reactions) are not easy to reverse. A burning piece of wood turns into a pile of ashes and gases that rise into the air. After the wood is burned, it cannot be returned to its original form as a log or twig. The following figure shows a chemical change in which hydrogen burns in oxygen to form water. A spark or flame will quickly combine hydrogen and oxygen into water chemically. But the reverse process, separating the elements hydrogen and oxygen from that water, is not as easy to achieve.

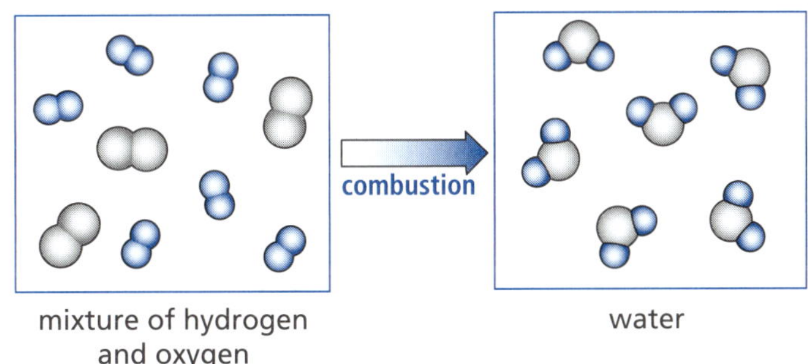

Even if gases or precipitates form, the law of **conservation of matter** still applies. This law states that matter is not created or destroyed in any chemical or physical change. When a piece of iron metal is heated in air, it combines with oxygen. (This happens as well when a piece of metal rusts.) If the resulting product is weighed, it has a greater mass than the original piece of iron metal. If, however, the mass of the oxygen that combines with the metal is taken into account, it can be shown that the mass of the product is equal to the sum of the masses of the iron and oxygen that combine.

iron + oxygen → iron oxide
total number of atoms or ions = total number of atoms or ions
total mass of reactants = total mass of products

There are ways to recognise that a chemical change has occurred.

- There is often a colour change.
- There is a release or gain of heat energy by the reaction. The reaction produces heat (gets hot) or takes in heat (gets cold).
- Light or sound energy may be released.
- There may be a change of odour.
- Bubbles of gas may form.
- A solid may precipitate from solution.

2 In a synthesis reaction, two or more reactants come together chemically to form one new compound. This can be written as A + B → AB. For example:

barium + oxygen → barium oxide

Notice that in the compound formed, the metal is named first followed by the non-metal with its ending changed to *ide*.

(cont.)

The arrow indicates that a chemical change has occurred. The following diagram represents a synthesis reaction.

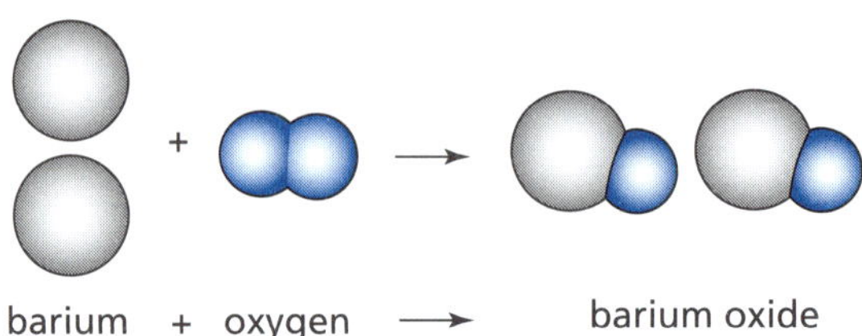

3 The reaction between a non-metal and a non-metal forms compounds which are molecular. In **molecular compounds**, the atoms share electrons. The reaction between a metal and a non-metal forms compounds which are ionic. In **ionic compounds** the metal atom gives up one or more electrons to the non-metal partner.

4 When one compound breaks apart into two or more atoms, it's known as a **decomposition** reaction. Sometimes the decomposition reaction occurs slowly over a period of time. An example is the decomposition of hydrogen peroxide, which slowly decomposes into water and oxygen.

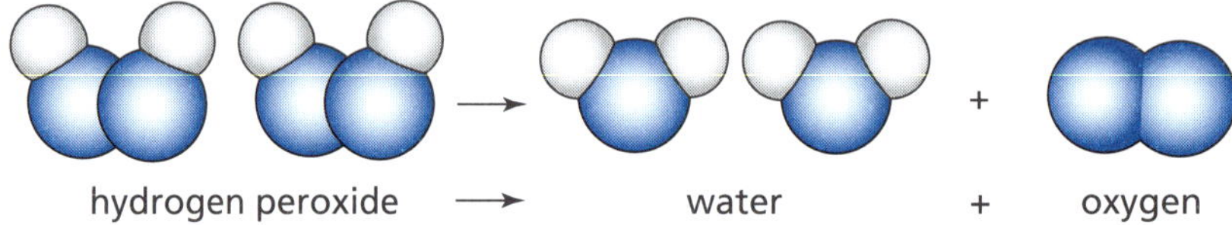

With other chemicals the reaction needs to be initiated, such as by heating. Other carbonates will decompose when heated, producing the corresponding metal oxide and carbon dioxide. For example, heating calcium carbonate leads to:

calcium carbonate → calcium oxide + carbon dioxide

Certain silver compounds are sensitive to light and decompose, depositing silver. This has allowed them to be used in black-and-white film photography. Decomposition reactions require a source of energy such as heat, light or electricity to decompose the compound involved. Even those substances that slowly decompose on their own, such as hydrogen peroxide, need to obtain enough energy before the decomposition can begin. For this reason, hydrogen peroxide should be stored in a cool, dark area.

Checklist

Can you:

1. *Distinguish chemical changes from physical changes and identify when a chemical reaction has occurred?* ☐
2. *Write simple word equations for synthesis reactions?* ☐
3. *Identify in synthesis reactions that non-metals + non-metals form molecular compounds and that metals + non-metals form ionic compounds?* ☐
4. *Identify and write simple word equations for decomposition reactions?* ☐

SIMPLE CHEMICAL REACTIONS

Elements and compounds

REVISION TEST

1 Consider the nursery rhyme:

Humpty Dumpty sat on a wall.
Humpty Dumpty had a great fall.
All the king's horses and all the king's men
Couldn't put Humpty together again.

a Could this be considered to be a chemical change or a physical change? Explain. *Hint 1* (2 marks)

b Many chemical reactions are not easy, or impossible, to reverse. Comment on this statement with reference to the nursery rhyme. (2 marks)

2 In a chemical reaction, the particles of one substance are rearranged to form a new substance. What does this statement say about the conservation of matter? (1 mark)

3 The diagram shows a burning candle.

a Is a physical change occurring at P or at Q? (1 mark)

b State three features that show that a chemical change is occurring when a candle burns. *Hint 2* (3 marks)

4 Complete the following synthesis reactions.

a iron + sulfur → ____________ sulfide (1 mark)

b ____________ + oxygen → magnesium oxide (1 mark)

c aluminium + bromine → aluminium ____________ (1 mark)

d sodium + iodine → ____________ ____________ (2 marks)

5 Ammonia gas has the formula NH_3. Write a word equation for the synthesis of ammonia. *Hint 3* (2 marks)

6 Carbonic acid, H_2CO_3, gives the fizz in soft drinks. Over time this will decompose spontaneously into carbon dioxide and water.

a Write a word equation for this reaction. (2 marks)

b Why does a glass of soft drink go flat if left for a few hours? (1 mark)

7 Complete these decomposition reactions.

a mercury oxide → mercury + ____________ (1 mark)

b magnesium ____________ → magnesium + chlorine (1 mark)

c ____________ ____________ → iron + sulfur (2 marks)

8 Decomposition can also split one compound into two simpler compounds (or a compound and an element).

a Consider this example: calcium carbonate → calcium oxide + carbon dioxide
Complete the reaction for the following:
sodium carbonate → ____________ ____________ + ____________ ____________ (2 marks)

b Consider this example: barium chlorate → barium chloride + oxygen
Complete the reaction for the following:
potassium chlorate → ____________ ____________ + ____________ (2 marks)

(cont.)

9 In an experiment a solution of sodium sulfate was placed in a test tube. This was inserted into a flask containing a solution of calcium chloride. The flask was stoppered. The set-up was carefully weighed before the reaction. The flask was tipped so the chemicals could react together. It was noticed that a precipitate forms. After the reaction was complete, the set-up was carefully weighed again.

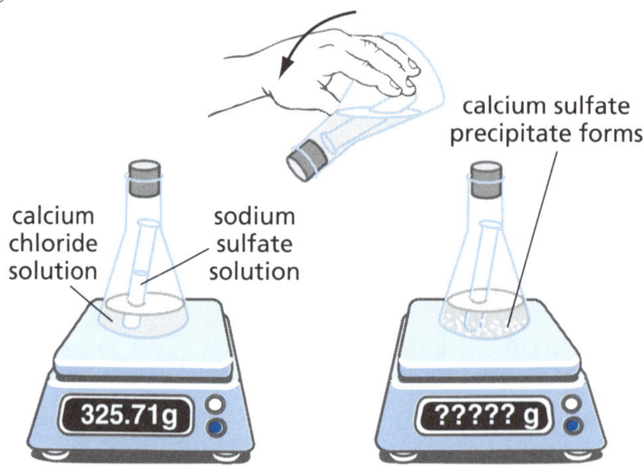

Will the weight after the reaction is completed be more, less or the same as before? Explain.
Hint 4 (2 marks)

10 When silver chloride is exposed to sunlight it slowly decomposes to form chlorine gas and silver.

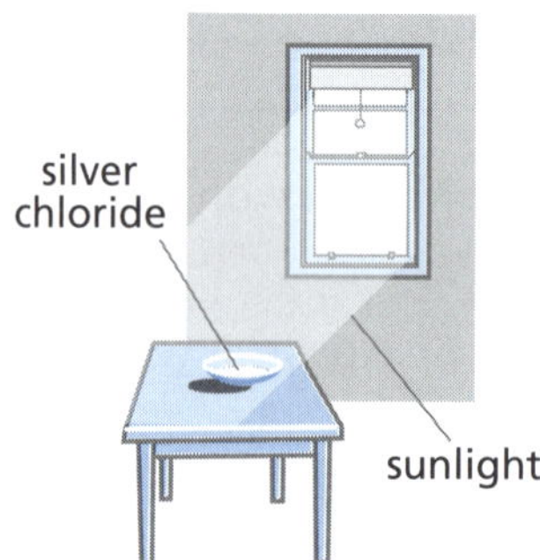

a Write a word equation for this reaction. (1 mark)

b As the reaction proceeds, the white-coloured silver chloride turns grey. Account for this observation. (1 mark)

c What happens to the chlorine gas produced? (1 mark)

Hint 1: Ask yourself whether or not a chemical reaction has occurred.
Hint 2: There are a number. Any three observable features will do.
Hint 3: The formula has been given as a hint. You do not need to use it in the word equation.
Hint 4: This container is sealed, so nothing can enter or leave.

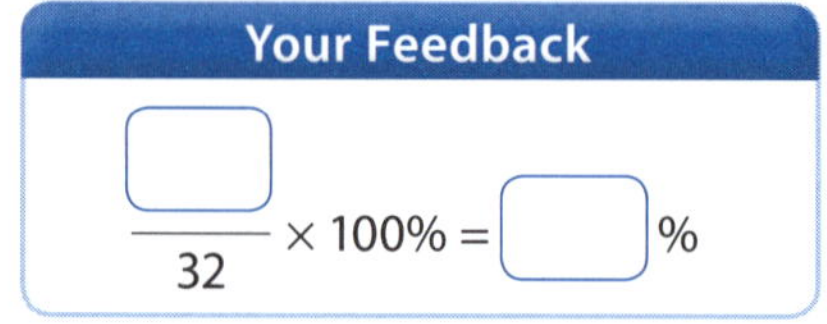

Strand: Earth and space sciences

EARTH STRUCTURE AND MINERALS

Minerals and rocks

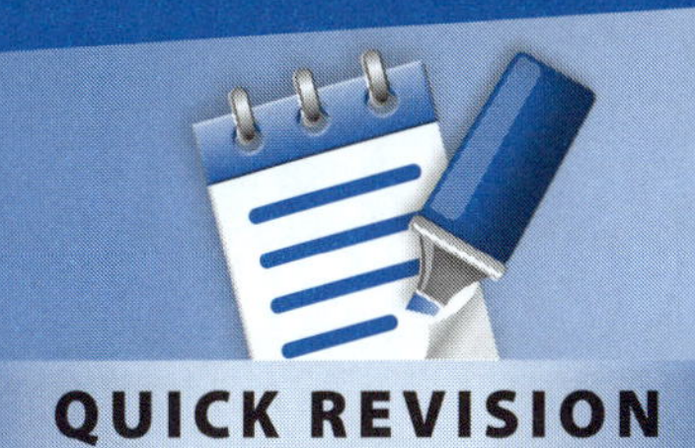

QUICK REVISION

1 The Earth is __________ as a rocky planet. For many years scientists had no idea about the internal structure and composition of the Earth. By analysing the waves produced by earthquakes as well as the shock waves caused by explosions, it was discovered that the Earth is composed of layers.

The ground on which we live is the top of the outer layer called the __________. The thickness of this rocky crust varies considerably. It is __________ under the continents and thinnest under the __________. The continental crust is about 35 km thick under mountain ranges and its average thickness is about 20 km. The oceanic crust is typically about 5 to 10 km thick. The crust is composed of a variety of __________. The most common group of minerals are classified as __________ which means that the predominant elements are silicon and oxygen. Different silicates contain various metals including __________, iron, calcium and potassium. Other minerals include sulfides (e.g. chalcopyrite, $CuFeS_2$) and oxides (e.g. haematite, Fe_2O_3).

The layer below the crust is the __________. It has the largest volume of all the layers (~80%). The top 100 km of the mantle is quite __________. Together with the crust this zone is called the __________. Below the lithosphere is a more mobile layer of the mantle called the asthenosphere.

2 Rocks are __________ of various minerals. A mineral is a naturally occurring inorganic solid compound that has a definite __________ composition. The elements in a mineral are arranged in an __________ structure called a crystal or crystal lattice. The most abundant elements by weight in the Earth's crust are __________, silicon, aluminium, iron, calcium, sodium and potassium. These elements are present in various common minerals such as quartz (silicon dioxide), orthoclase feldspar (potassium aluminium silicate) and plagioclase feldspar (calcium sodium aluminium silicate). These minerals have different __________ structures.

Minerals can be identified using their physical and chemical properties. Crystals have a variety of __________ shapes. Salt crystals, for example, are cubic in shape. Some crystals are octahedral, rhomboidal or needle-like. The cleavage of a mineral refers to its tendency to break along specific smooth crystal __________ when struck by a hammer.

Minerals also vary in their hardness. The hardness of a mineral is expressed on a 10 point scale called the Mohs scale of hardness. Diamond is the __________ natural mineral (hardness = 10). Talc minerals are the softest (hardness = 1). The common mineral, quartz, has a hardness of 7.

The colour of a mineral may be __________ due to impurities. Pure quartz is colourless whereas traces of titanium or iron in the quartz can produce rose quartz which has a pink hue. Citrine is a variety of quartz that is typically yellow. The colour of a mineral may also change when the mineral is __________ to a powder. The colour of the powder is called its __________. Streak colours are fairly constant even if impurities are present in a mineral. Calcite is a common mineral which is white or colourless if pure but it can be many different colours when impure. Its streak, however, is always white.

Minerals have a variety of different chemical properties. Carbonate minerals such as calcite ($CaCO_3$) and magnesite ($MgCO_3$) both __________ when drops of acid are added to them. This fizzing is caused by the production of carbon dioxide gas.

Answers **1** classified; crust; thickest; ocean; minerals; silicates; aluminium; mantle; rigid; lithosphere **2** mixtures; chemical; ordered; oxygen; crystalline; geometric; planes; hardest; variable; ground; streak; fizz

EARTH STRUCTURE AND MINERALS

Minerals and rocks

REVISION SUMMARIES

1 The Earth consists of four layers. The outermost layer of the Earth is called the **crust**. On average, the crust is about 20 km thick. This is quite a small distance compared to the Earth's radius (~6370 km). On the continents the crust is about 35 km thick, but it is thinner under the oceans, roughly 5 to 10 km in thickness. Rocks in the Earth's crust have an average density 2.8 times that of water. This is much less dense than other layers. In terms of chemical composition, the Earth's crust is a mixture of various minerals. Silicate minerals (containing the elements silicon, oxygen, aluminium, iron and calcium) make up about 92% by weight of the crust.

The temperature of the crust gradually increases with depth. It reaches values of about 1000 °C at the base of the crust where it meets the next layer called the mantle. Over 80% of the Earth's volume lies in the mantle. The **lithosphere** is the rigid upper part of the Earth. It consists of the crust and the rigid part of the upper mantle. The parts of Earth's structure are shown in the diagram below.

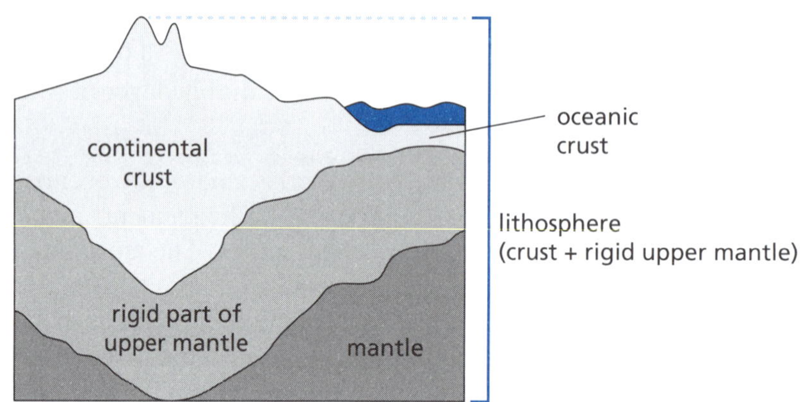

2 Most rocks on the Earth's surface are made up using only eight elements: oxygen, silicon, aluminium, iron, magnesium, calcium, potassium and hydrogen. But these elements can be combined in many different ways to produce minerals that are very different. A **rock** is made of one or more minerals. The minerals are found together in the rock as a mixture.

A **mineral** is a naturally occurring inorganic solid compound that has a definite chemical composition. The atoms in a mineral are arranged in an ordered array. This orderly stacking is easily seen in the regular shape of the crystal. Geologists have developed a number of physical methods to quickly identify minerals. Here are some of these physical properties.

- **Crystal shape.** Some crystals have a very distinctive crystal shape with well-formed crystal faces. Shapes are based on common geometric shapes such as cubic, rhombic and hexagonal.
- **Lustre.** The surfaces of crystals may be shiny, dull, metallic or pearly.
- **Hardness.** Diamonds are very hard. They can scratch glass and this is why glaziers use diamond-tipped glass cutters for this purpose. Minerals can be arranged in order from hardest to softest using **Mohs scale** (created in 1812 by the German geologist and mineralogist Friedrich Mohs). This scale of mineral hardness is based on the ability of one natural sample of matter to scratch another. Diamonds are rated as 10 on this scale, as they were the hardest-known naturally occurring substances when the scale was designed. The scale is not linear. For example, corundum (9) is twice as hard as topaz (8), but diamond (10)

is four times as hard as corundum. However, it does give a relative indication of the hardness of different minerals.

You can find how hard a mineral is by rubbing the mineral of unknown hardness against one of known hardness. The one which does the scratching is harder than the object being scratched. The following table outlines the Mohs scale of hardness.

Mohs scale	1	2	3	4	5	6	7	8	9	10
Mineral	talc	gypsum	calcite	fluorite	apatite	feldspar	quartz	topaz	corundum	diamond

- **Colour.** Colour may not be a reliable characteristic due to impurities in the mineral. For example, the mineral quartz can appear in a wide variety of colours from clear to black. Purple quartz, for instance, is known as amethyst while rose quartz is pink and translucent.
- **Streak.** This is the colour of the mineral in its powdered form. This colour is found when a sample of the mineral is rubbed across a streak plate (a tile of unglazed porcelain). While the colour of a mineral can vary, its streak usually does not.
- **Cleavage.** This is the tendency of a mineral to break along planes of weak bonding. These are called cleavage planes and they are smooth.

The following diagram shows the crystal shapes of some common minerals. These varying shapes help to identify minerals.

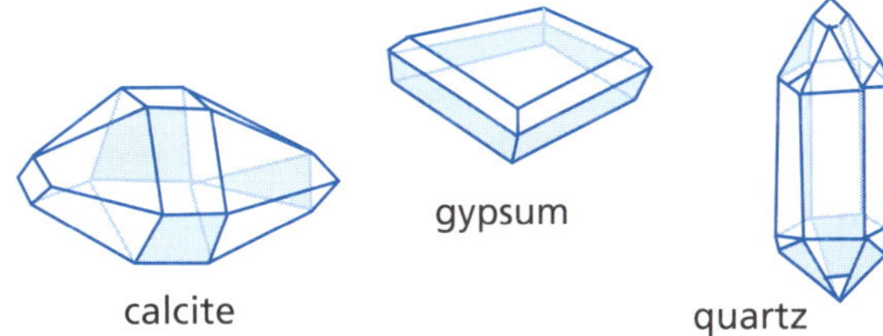

Chemical properties may also be useful in identifying minerals. For example, calcite minerals are composed of calcium carbonate. Carbonate minerals react with dilute acid and release carbon dioxide gas. When drops of hydrochloric acid are added to the surface of calcite or a rock containing calcite, there is a fizzing reaction as the gas (carbon dioxide) is released.

Checklist

Can you:

1 *Describe the structure of the lithosphere?* ☐

2 *Define the term 'mineral' and describe the properties of a mineral that allows it to be identified?* ☐

EARTH STRUCTURE AND MINERALS

Minerals and rocks

REVISION TEST

1 **a** Your fingernail can scratch the mineral gypsum. Which is harder, your fingernail or gypsum? (1 mark)

b Quartz is scratched by a steel file. Which is harder? (1 mark)

c The mineral calcite can scratch your fingernail, but will not scratch glass. Arrange these in order from softest to hardest. (1 mark)

d Read the following information and arrange the substances in order from least hard to most hard. (1 mark)

- A steel file can scratch a pocket knife blade.
- Calcite scratches a copper coin.
- A pocket knife scratches calcite.
- Quartz can leave a mark on a steel file.
- Quartz is scratched by a topaz.

e Haematite is a mineral containing the compound iron oxide. Its colour is very dark, and may be almost black, dark red or brown. Its streak, however, is blood red.

i How is this colour indicated in the name for the mineral? *Hint 1* (1 mark)

ii Why is the streak a better indicator for this mineral? (1 mark)

2 Density is an important property of minerals. It is calculated using this formula:

$$\text{density} = \frac{\text{mass}}{\text{volume}} \quad \text{or} \quad D = \frac{m}{V}$$

a Ten cubic centimetres of quartz has a mass of 27 g. What is the density of quartz? (1 mark)

b Chalcopyrite is an important mineral ore of copper. Its colour is so golden it is often mistaken for gold. This is why it is sometimes called fool's gold. Chalcopyrite has a density of 4.2 g/cm^3, and gold has a density of 19.3 g/cm^3. A golden mineral sample had a volume of 8.0 cm^3 and a mass of 33.6 g. Is this mineral gold or fool's gold? (2 marks)

3 The table below lists the properties of common minerals. Use the data to answer the questions that follow.

Mineral	Colour	Streak	Hardness	Lustre
apatite	w, gr, b, y, p	w	4.5–5	r to v
calcite	c, w, b	w	3	d to v
chalcopyrite	y	gr, bl	3.5–4	m
feldspar	w, pink	c	6	p to v
fluorite	w, y, gr, r, p, b	w	4	v
galena	g	g	2.5	m
gypsum	w, c, g, y, r	w	1.5–2	p to v
haematite	g	r, br	5–6	m
muscovite	gr, br, y, c	c	2.7–3	v
olivine	gr, br	c	6.5–7	v
pyrite	y	gr	6–6.5	m
quartz	c, p, r, y	w	7	v
sphalerite	y, br, r, bl, w	w	3.5–4	r to b
talc	w, gr	w	1	p

Colour and streak:
b = blue
bl = black
br = brown
c = colourless
g = grey
gr = green
p = purple
r = red
w = white
y = yellow

Lustre:
b = brilliant
d = dull
m = metallic
p = pearly
r = resinous
v = vitreous (like glass)

a Identify the property that is most useful in distinguishing between chalcopyrite and pyrite. (1 mark)

b Name the mineral that has the following properties: colour is grey, streak is grey, hardness is 2.5 and lustre is metallic. (1 mark)

c Identify a white mineral in the table that will scratch haematite. *Hint 2* (1 mark)

4 a Name the outermost layer of the four layers of the Earth. (1 mark)

b Name the rigid part of the Earth. (1 mark)

c Name the two most common elements in the crust of the Earth. (1 mark)

d A mineral was extracted from the crust and tested with drops of hydrochloric acid. The mineral did not fizz. What conclusion could you make about this mineral? *Hint 3* (1 mark)

e What is the hardness of calcite on Mohs scale? (1 mark)

f True or false? Oceanic crust is thicker than continental crust. (1 mark)

g True or false? The mantle is more rigid than the lithosphere. (1 mark)

5 The following data shows the elemental percentage composition of the crust by weight for the five most abundant elements.

Element	oxygen	silicon	aluminium	iron	calcium	others
Composition of crust (%)	46.6	27.7	8.2	5.0	3.6	

a Calculate the percentage by weight of the remaining elements (others). (1 mark)

b Use this data to construct a pie (sector) graph. (6 marks)

Hint 1: What is the name of the red oxygen-carrying molecule in red blood cells?

Hint 2: A harder mineral will scratch a softer mineral.

Hint 3: No carbon dioxide gas formed. What minerals release carbon dioxide when treated with acid?

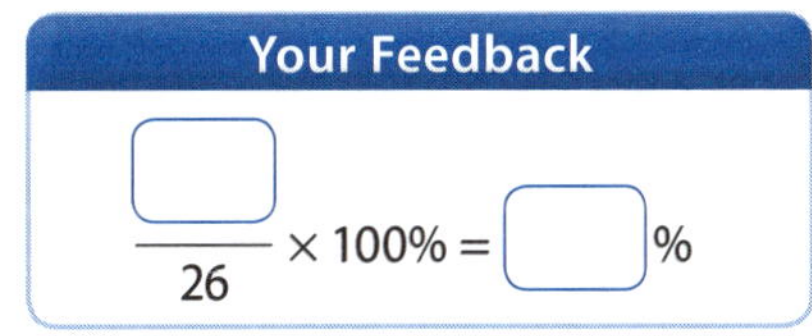

PAGES 190–191
PAGE 212

WEATHERING AND EROSION

Minerals and rocks

QUICK REVISION

1 The natural landscape __________ over time. Forces inside the Earth cause mountain building but this is __________ by other forces that wear away mountains. Sand dunes are in a state of constant motion due to the action of __________ and rain. Fast-flowing rivers and glaciers can also __________ away the land and create new landscapes.

The Earth's crust is made of rocks of varying types and mineral compositions. These rocks are changed by a process called weathering. One type of weathering is called __________ weathering. In this process the rock is broken down into smaller pieces. The minerals in the rock are __________ changed. A second types of weathering is called chemical weathering. In this process chemical changes occur which __________ the minerals permanently.

2 Physical weathering can occur in many different ways. One common example is due to repetitive temperature changes over many years. During the day exposed rocks __________ heat from the Sun and their surface layers __________. At night these surfaces cool down and __________. Over many years, this process leads to the formation of __________ and a destabilisation of the surface. The photo shows cracks in rock that have formed by repetitive heating and cooling.

Cracks in the surface layers of rock can be widened in alpine areas. If water becomes trapped in these cracks in the winter, the water will __________ to form ice as the temperature drops below 0 °C. Ice takes up more space than liquid water as it is __________ dense. The expansion of water to form ice leads to great pressure being applied to the crack. The existing crack will get __________ as a result and new cracks will develop.

Rock cracks can also fill with soil and this encourages the growth of seeds that may fall into the soil. As a plant grows in these cracks, its __________ exert great forces which lead to further cracking of the rock.

3 Chemicals in the environment can also produce chemical weathering of rocks. Some rocks are more susceptible to chemical weathering than others. Feldspar minerals, for example, chemically weather and are __________ into clay minerals. Clay minerals swell in contact with water and become softer. Such changes are called __________. Carbonate minerals are also subject to weathering by hydration.

Oxygen in air and water can chemically weather various minerals and rocks. For example, chalcopyrite minerals ($CuFeS_2$) are __________ to produce iron oxides and copper hydroxides.

WEATHERING AND EROSION

Minerals and rocks

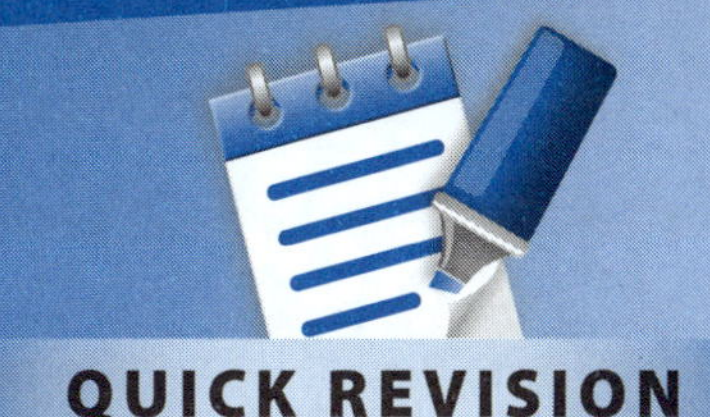

QUICK REVISION

There are many weak acids in nature and these chemicals can attack minerals and rocks. One common acid in the environment is ___________ acid which forms when carbon dioxide dissolves in water. Limestone rocks are composed of the mineral called ___________. Carbonic acid can attack and dissolve calcite with the release of carbon ___________ gas. Pollution of the environment can also lead to the production of acidic solutions. Acids such as ___________ acid can be formed when sulfur dioxide gas (produced during the smelting of sulfide ores) reacts with ___________ and oxygen. Sulfuric acid can convert the marble in a statue into a new mineral called gypsum which is much softer.

4 Once a rock has been weathered, the weakened material is readily ___________. Erosion refers to the ___________ of weathered material away to a new location. One common agent of erosion is gravity. A landslide is an example of ___________ caused by gravity. The following photo shows a house buried by a landslide.

Moving water is also a common agent of erosion. Fast-flowing rivers can ___________ the banks of a river and then transport ___________ material downstream. The sediment that is transported by the river will be ___________ in calmer water such as a lake or the sea. Ocean waves apply great forces against cliffs on the coast. Loose material is then carried out into ___________ waters. Torrential rain and floods can carry away ___________ and other weathered materials.

Ice in the form of a glacier is also an agent of erosion. Glaciers erode the valleys in which they are located. Gravity causes the glacier to move ___________ and the glacier carves out U-shaped valleys as it abrades the rocks over which it slides. Large pieces of ___________ can be transported downhill by this process. Wind is a common agent of erosion. In a desert, wind can pick up ___________ and use it to blast the surfaces of rocks and abrade them. On the shoreline, ___________ sand can blast the surfaces of the cliff-face and wear away the cliff.

Answers **1** changes; opposed; wind; wear; physical; not; alter (change) **2** absorb; expand; contract; cracks (crevices); freeze; less; larger; roots **3** converted (turned); hydration; oxidised; carbonic; calcite; dioxide; sulfuric; water **4** eroded; transport; erosion; abrade; loose; deposited; quieter (calmer); soil; downwards; rock; sand; windblown

WEATHERING AND EROSION

Minerals and rocks

1 All types of **landscape** change with time. Rivers change their direction, landslides occur in the mountains and coastal sand dunes are destroyed during storms. The surfaces of rocks are constantly attacked by environmental forces that lead to permanent changes.

Weathering refers to the changes that lead to the breakdown of the Earth's crust. Weathering can be classified in two main ways.

- **Physical weathering** involves the breakdown of rocks into smaller and smaller pieces without any change to the mineral composition of the rock.
- **Chemical weathering** involves a permanent alteration in the chemical composition of the minerals of which the rock surface is composed. The rock surface begins to crumble.

2 The following are some examples of physical weathering.

- **Temperature changes.** During the day, the exposed surfaces of rocks heat up and expand. At night the rocks cool and the surface contracts. This repetitive process leads to a strain in the rock and ultimately the surface cracks.
- **Root action.** Plants grow in the soil that collects in rock crevices. As their roots grow larger, the force they exert causes the crevices to grow larger.
- **Ice cracking (frost wedging).** Ice cracking only occurs where the temperature drops below freezing point (0 °C). During a very cold night, the water trapped in small rock **crevices** freezes. As the water freezes, it expands and exerts a great pressure in the rock. Existing cracks widen as shown in the following diagram. In Australia, this type of weathering tends to occur only in alpine areas. Along the coast, the temperatures are mild and ice cracking is rare.

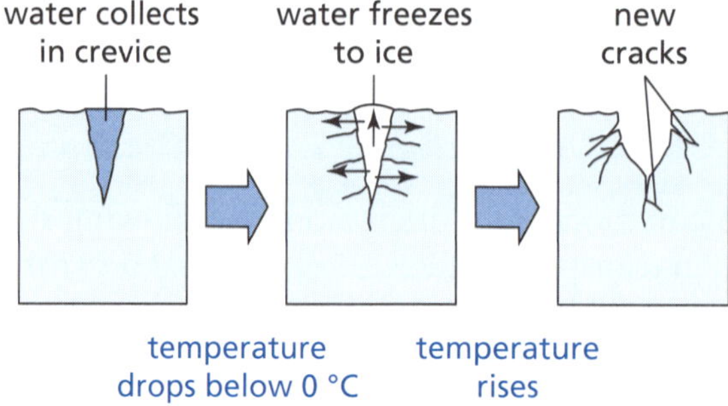

3 The following are some examples of chemical weathering.

- **Oxidation.** Oxygen in the air or oxygen dissolved in water can attack rock minerals and convert them into new, softer substances which tend to be crumbly. Rocks that contain iron minerals turn a rusty brown as this chemical weathering occurs. If you break open a weathered rock, you will see that the unweathered parts are quite a different colour.
- **Water (hydration).** Rocks that contain clay minerals are affected by water. The clay absorbs the water and swells as it changes its structure and becomes softer. As the clay dries, it contracts. This repetitious swelling and contraction leads to weathering.
- **Acids.** Living things such as plants, algae and lichens which grow near or on rocks produce acidic juices that slowly attack the rocks. Carbon dioxide in the air dissolves in water to produce a weak acid called **carbonic acid**. This acid particularly attacks and dissolves limestone rocks. Sinkholes and caves can be created by the action of carbonic acid. These landscape features are now called karst topography. Pollution can cause much stronger acids (sulfuric and nitric acid) to form in rain. The rain chemically attacks rocks, buildings and statues.

4 **Erosion** is the wearing down of the Earth's crust and the carrying away of weathered material and other sediment to different locations. The weathered rocks are quite susceptible to agents of erosion. Some of these agents are:

- moving water such as rain, rivers and waves
- gravity, which pulls loose material downhill, such as landslides
- wind, which picks up sediments that are then rubbed against other rocks; sand particles can be picked up by the wind and then further abrade rocks along the coast or in a desert
- moving ice such as glaciers, which carve out U-shaped glacial valleys and transport rocks and pebbles down the valleys.

The action of moving water is by far the most common cause of erosion. As rivers rush down from the mountains towards the coastal strip, they erode the surrounding land and transport the eroded material (the sediment) in mostly suspended form. The sediment is **deposited** in calm bodies of water such as lagoons, estuaries or lakes, or in the shallow coastal seas over the continental shelf. Deposition of sediment also occurs near river bends as the water changes direction and slows down as shown in the diagram below. Underwater currents can carry the sediment into deeper ocean waters. Fast-flowing streams can transport sand, pebbles and larger sediments further than slow-moving streams which mainly transport fine clay and mud particles.

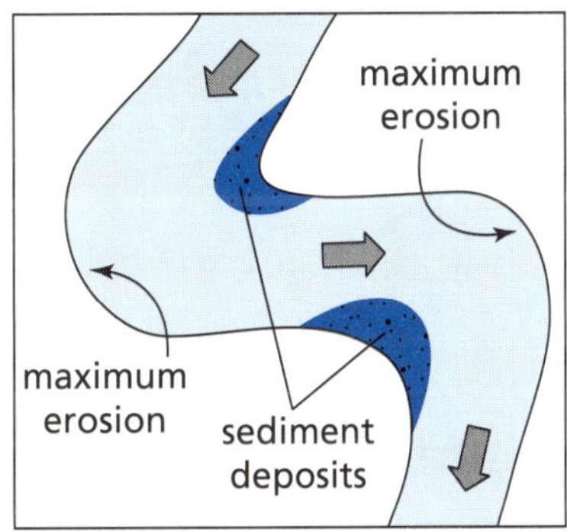

Checklist

Can you:

1. *Define the term 'weathering'?* ☐
2. *Give examples of physical weathering?* ☐
3. *Give examples of chemical weathering?* ☐
4. *Define 'erosion' and name various agents of erosion?* ☐

WEATHERING AND EROSION

Minerals and rocks

REVISION TEST

1 The diagram below shows an example of erosion called soil creep. Identify the agent of erosion. (1 mark)

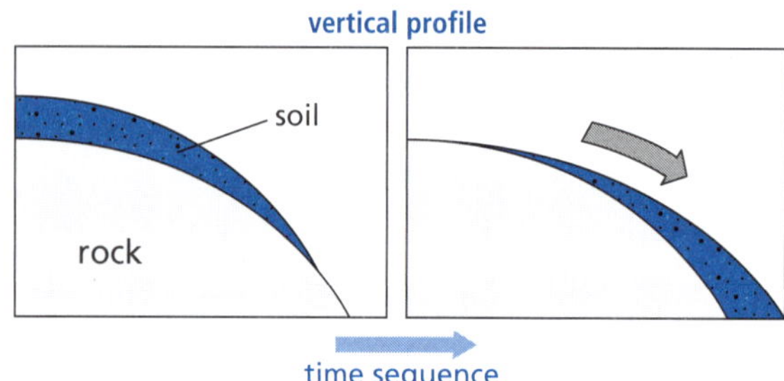

2 The following diagram shows a house built near the beach. The owner decided to build a sea wall out of rocks to protect his home from wave action. He was afraid that the supports of the house would be undermined by beach erosion.

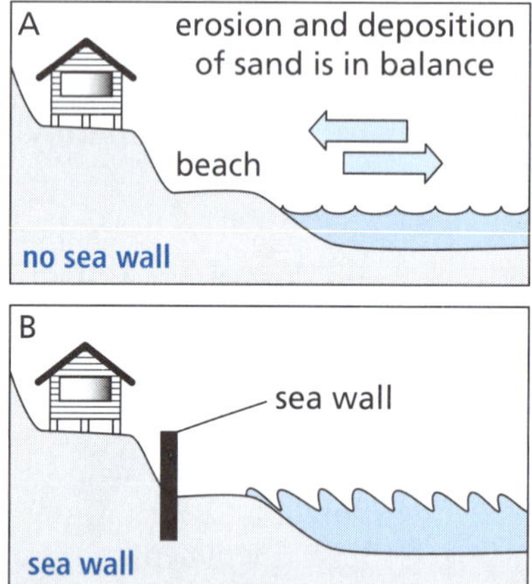

Predict whether or not the erosion and deposition of sand will still remain in balance after the construction of the sea wall. Explain your answer. *Hint 1* (2 marks)

3 State whether the following statements are true or false.

- **a** Wind is the major agent of erosion. (1 mark)
- **b** Ice cracking is an example of physical weathering. (1 mark)
- **c** Streams carry sediment towards calm bodies of water. (1 mark)
- **d** The widening of rock crevices by root action is an example of chemical weathering. (1 mark)
- **e** Along the coast, erosion is mainly caused by wind action. (1 mark)

4 Gibber deserts in central Australia are covered with stones with polished surfaces.

- **a** What agent of erosion is likely to have caused this effect? (1 mark)
- **b** Can this type of erosion occur on the coastal strip? Explain. (2 marks)

5 Coastal rivers carry rock material towards the sea. The river carries the material in several ways:

- materials dissolved in the water
- materials suspended in the water
- materials rolled and pushed along the river bed.

Classify each of the following materials into one of the three groups.

a pebbles (1 mark)
b mineral salts (1 mark)
c clay (1 mark)
d boulders (1 mark)
e fine sand (1 mark)
f very coarse sand (1 mark)

6 Students performed an experiment in which they used a pair of tongs to hold a small piece of sandstone rock in the hottest part of a Bunsen flame for 2 minutes. They then dipped the hot rock into a dish of cold water. This procedure was repeated six times.

a What safety measures must students take in this experiment? (3 marks)
b What type of weathering are students investigating? *Hint 2* (1 mark)
c What observation will show that the weathering process has occurred? (1 mark)

7 A student conducted an experiment at home using a small piece of limestone rock and soda water. Her aim was to determine whether the soda water could chemically weather the limestone over a period of several months.

a Explain why the student chose soda water for this experiment. *Hint 3* (1 mark)
b What could she measure that would show weathering had occurred? (1 mark)
c How could she improve the reliability of her experiment? (1 mark)

8 The following diagram shows a sea cave and a karst sinkhole that have formed on an island. Explain how these features are formed in limestone rocks. (2 marks)

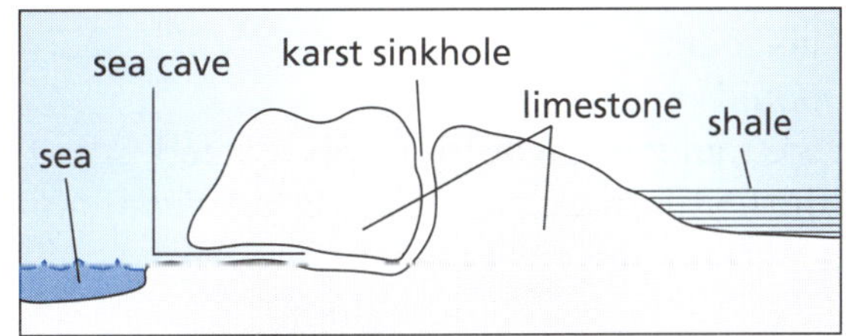

Hint 1: As long as sand continues to be deposited on the beach at the same rate as erosion occurs then the beach will remain stable. If erosion exceeds deposition then the beach will wear away.
Hint 2: In nature the Sun provides the heat.
Hint 3: What chemical is dissolved in the water during the manufacture of soda water?

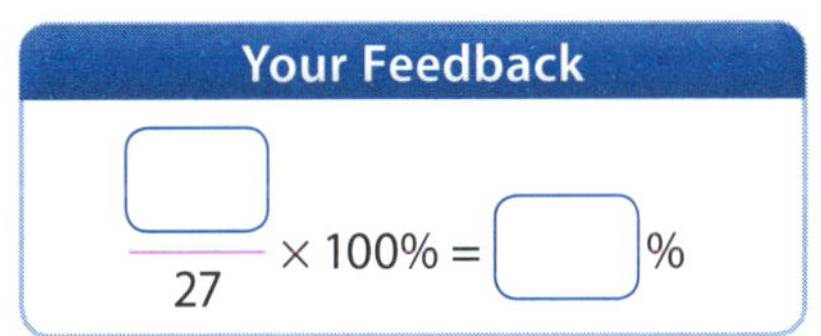

SEDIMENTARY ROCKS AND FOSSILS

Minerals and rocks

QUICK REVISION

1 For many thousands and even ________ of years, rock material on our Earth has been ________; that is, broken down and worn away by wind, ice and water. These particles, some larger than others, may be washed downstream by ________, eventually settling to the bottom of the rivers, lakes or ________. In some areas very fine silt and sand might be transported long distances by ________. Layer after layer of the eroded earth surface is ________ one on top of the other. These layers are ________ down more and more over a long period of time, until the bottom layers slowly become ________.

There are several steps involved in this process after the rocks have been eroded.

- The material is ________ to a new site where it will settle.
- As flowing streams and rivers ________, the material is dropped and sinks to the bottom.
- Over millions of years, ________ build on top of each other, putting pressure on lower layers.
- The ________ squeezes out water and allows ________ material to bind the fragments together.
- In the meantime, the ocean, lake or other body of water ________ up. This makes the sedimentary layer, which used to be under water, become a layer in dry ground.

Different rock layers can be different if erosion drops different things to the bottom of the water for each layer. This accounts for streaks of different ________ or textures in sedimentary rock.

2 There are many examples of sedimentary rocks.

- ________ is made from small grains of the minerals quartz and feldspar. Sandstones are often used in buildings and are commonly seen in those built over a century ago.
- Conglomerate consists of large sediments such as sand and ________ or other rounded fragments. The sediment is so large that pressure alone cannot hold the rock together; it is also ________ together with dissolved minerals.
- Gypsum contains calcium sulfate mineral and is formed by sea water ________ in huge prehistoric basins. This sedimentary rock is very ________ and is used to make plaster of ________, casts, moulds and wallboards.
- Limestone contains the mineral calcite, and was formed from the beds of evaporated seas and lakes and from the fossils of ________. It most commonly forms in clear, warm, shallow ________ waters. This rock makes a very good building stone for humid regions and is also used to make ________ for building houses.
- Shale is formed from clay particles that are compacted together by ________. This rock can be used to make bricks.

3 Fossils form in several ways. But in all cases the environmental ________ need to be just right for hard or even soft parts to survive being buried deep in ________ for millions of years. Fossils are often found in sedimentary rocks and the plant or animal material becoming the fossil develop as the sediments are ________ around it to produce these rocks. Some dead organisms are buried in the right ________ place and conditions so they can become fossils. The location and conditions at the burial site have to be just right for a fossil to be formed. The study of fossils, how they were ________ and the evolutionary relationships between animal groups are important functions of the science of palaeontology.

Answers **1** millions; eroded; rivers; oceans; wind; deposited; compressed (weighted); rock (hard); transported (carried); slow (reduce speed); layers; pressure; cementing (mineral); dries; colours **2** sandstone; pebbles; cemented; evaporating; soft; Paris; seashells; marine; cement (concrete); pressure **3** conditions; sediment (rock); hardening; places (location); formed (created, produced)

SEDIMENTARY ROCKS AND FOSSILS

Minerals and rocks

REVISION SUMMARIES

1 **Sedimentary** rocks are formed through the deposition and **solidification** (hardening) of **sediments**, especially sediment **transported** by water (rivers, lakes and oceans), ice (glaciers) and wind. Sedimentary rocks are often deposited in layers and may contain fossils.

Sedimentary rocks are formed from weathered or broken-down rocks or from the remains of once-living organisms. They originate from deposits that accumulate on the Earth's surface. There are several types of sedimentary rocks.

- **Clastic sedimentary rocks** are made up of pieces (clasts) of pre-existing rocks. Pieces are loosened by weathering and are then transported (eroded) to some basin or depression where the sediment is trapped. As the sediment gets buried more deeply, it becomes compacted and cemented forming sedimentary rock. Particle sizes can vary from microscopic clay to huge boulders.
- **Biological sedimentary rocks** form when large numbers of living organisms die, pile up and become compressed and then cemented to form rock. Accumulated carbon-rich plant material may form coal. Deposits made mostly of animal shells can form limestone or chert.
- **Chemical sedimentary rocks** are formed by chemical precipitation. The stalactites and stalagmites in limestone caves form like this, as does rock salt. The process begins when water travelling through rock dissolves some of the minerals and carries them away from their source. Eventually these minerals may be re-deposited, or precipitated, when the water evaporates or when it becomes over-saturated with minerals.

The following diagram outlines the steps in the formation of a sedimentary rock, showing transportation, sedimentation, compaction and cementation.

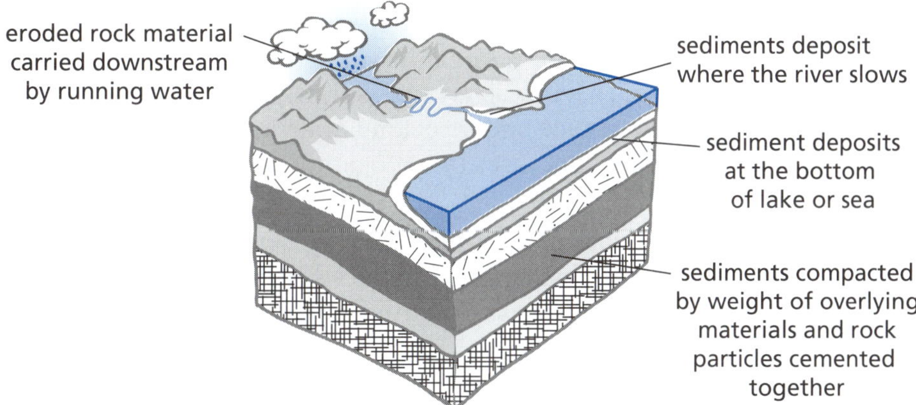

Cementation is the hardening and welding together of clastic sediments (those formed from pre-existing rock fragments) when mineral matter precipitates in the pore spaces. Pore spaces within a rock body are unoccupied by solid material. Porosity is a measure of the empty spaces in a rock. Well-sorted materials, with grains of about the same size, have higher porosity than similarly sized but poorly sorted materials. This is because smaller particles fill the gaps between larger particles. When the empty pore space becomes filled with solids, it makes the rock harder. A variety of minerals can become cements, such as silica (generally quartz), calcite and other carbonates, as well as iron oxides and clay minerals. These cementing minerals may come with the rock fragments, or be brought in from outside by circulating waters. This cement forms an integral and important part of the rock, and is the last stage in the formation of a sedimentary rock.

(cont.)

SEDIMENTARY ROCKS AND FOSSILS *(continued)*

Minerals and rocks

2 Clastic sedimentary rocks are named according to their **clast** or grain size. The smallest grains are clay, then silt, then sand. Larger grains are known as pebbles. Shale is a rock made mostly of clay, siltstone is made up of silt-sized grains, sandstone is made of sand-sized clasts and conglomerate is made of pebbles surrounded by a matrix of sand or mud. The following table lists examples and features of sedimentary rocks.

Rock type	Example	Features
clastic	shale	mostly clay minerals; compact; may easily split
	siltstone	very fine grains of silt
	sandstone	fine to coarse sand grains
	breccia	angular fragments; contains pebbles, cobbles and/or boulders in a matrix of clay, silt and/or sand
	conglomerate	rounded fragments; contains pebbles, cobbles and/or boulders in a matrix of clay, silt and/or sand
biological	coal	compacted plant remains; mostly carbon
	limestone	precipitates from biological origin or broken-down shell fragments; mostly calcite
chemical	rock salt	salt crystals from evaporates and precipitates

3 Sedimentary rocks sometimes contain fossils. **Fossils** are any remains, impression or trace of a living thing of a former **geological period**, such as a skeleton, footprint or leaf imprint, that is embedded and preserved in the Earth's crust. The location and conditions have to be just right for a fossil to form. For example, a potential fossil needs to be buried quickly so that it is not attacked by scavengers or the weather. Then it must be in an area where rock will form around it, so that the plant or animal's body will be **preserved**. The fleshy parts of buried animals and plants are normally eaten by small organisms or slowly decay away. This leaves the hard body parts like bones and teeth. In areas where the ground is slowly sinking, the skeleton is buried more and more deeply below the surface under accumulating layers of sediments. Increasing **pressure** on the lower layers of sediment fuses them together and results in a sedimentary rock. Hard body parts may stay intact within the rock.

Minerals may be deposited in the spaces within the bone by circulating water and makes the bones stronger. The water may also dissolve away minerals from bones. Fossils deep in the Earth need to be lifted back up to the surface. Eventually the ground moves upwards and the overlying material is eroded away to expose the fossil. Fossils that have been completely turned to stone are called **petrified fossils**. These are rock-like and much stronger than the original organic material. Petrified fossils show the internal structure of bones, teeth and shells, and even tree trunks have fossilised as petrified wood. Not all fossils are formed from bodies or body parts. **Impressions** such as footprints in mud can become fossils if different sediments fill and cover the footprints to protect them. Preserved evidence of an organism's activities, like footprints, burrows, eggs or nests, are known as trace fossils.

Checklist

Can you:

1 *Explain how sediments form sedimentary rocks, and recognise the role of transportation, sedimentation, compaction and cementation?* ☐

2 *Name some sedimentary rocks and how they form?* ☐

3 *Describe how fossils form and give simple examples?* ☐

SEDIMENTARY ROCKS AND FOSSILS

Minerals and rocks

REVISION TEST

1. Insert the missing word to complete each sentence.
   - **a** An example of a biological sedimentary rock is ____________. (1 mark)
   - **b** In the processes responsible for changing sediments into sedimentary rock, the step after compaction is ____________. (1 mark)
   - **c** Coal is formed from ____________ material. (1 mark)
   - **d** A rock composed of cemented, rounded pebbles is a ____________. (1 mark)
   - **e** Remains or traces of ancient plants and animals that are preserved in rock are called ____________. (1 mark)
   - **f** Breccia is an example of a ____________ sedimentary rock. (1 mark)
   - **g** Shale refers to a rock formed from ____________ minerals. (1 mark)
   - **h** Most shells of marine organisms are composed of ____________ carbonate. (1 mark)
   - **i** Sediments come from the ____________ of rock. (1 mark)
   - **j** Sedimentary rocks provide clues about the surface conditions at the time the sediment was ____________. (1 mark)

2. The following diagram shows the steps in forming sediments and sedimentary rocks. Write captions to replace the letters P, Q, R and S. (4 marks)

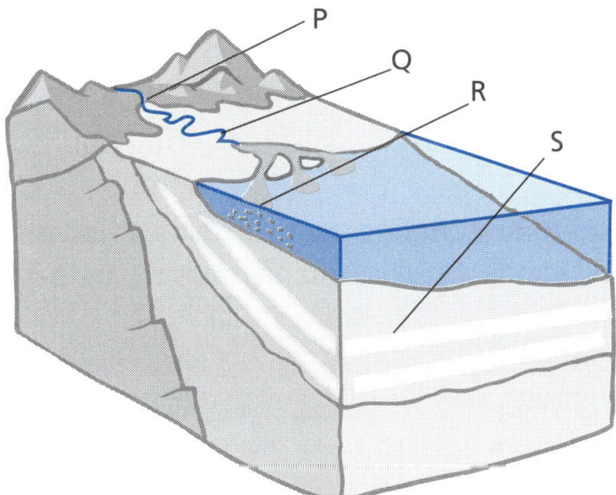

3. Many sedimentary rocks, such as mudstone and shale, show a typical layered structure. Why is this? *Hint 1* (1 mark)

4. The following diagram shows a microscopic view of a sedimentary rock.

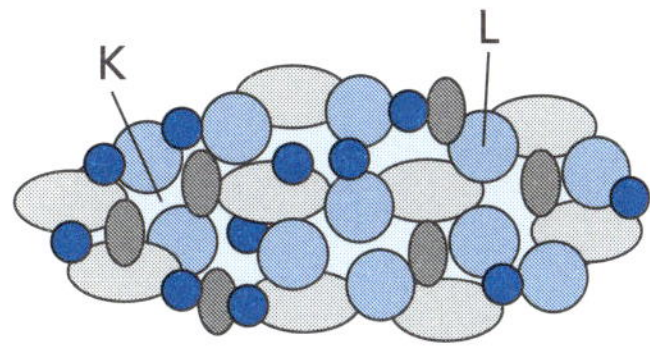

   - **a** What is represented by the shape labelled L? (1 mark)
   - **b** How many different kinds of particles are shown here? (1 mark)
   - **c** What is represented by the irregular shape labelled K? *Hint 2* (1 mark)
   - **d** Explain the purpose of the material in K. (1 mark)

(cont.)

5 Fossilisation is quite a rare event since only a small proportion of the creatures that die are ever preserved. The panels in the diagram below show an example of how fossils form. Write captions for each panel. (5 marks)

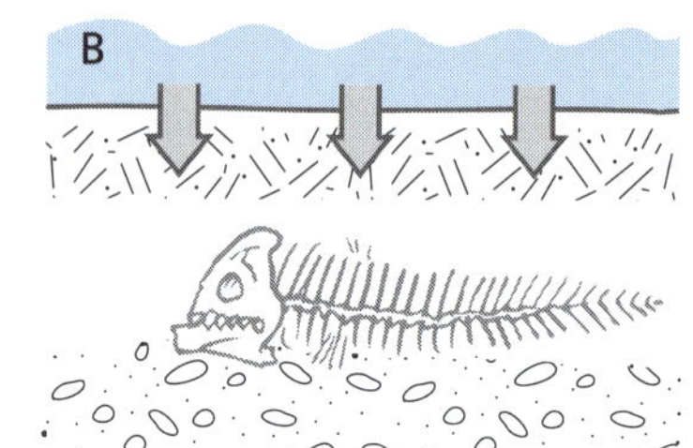

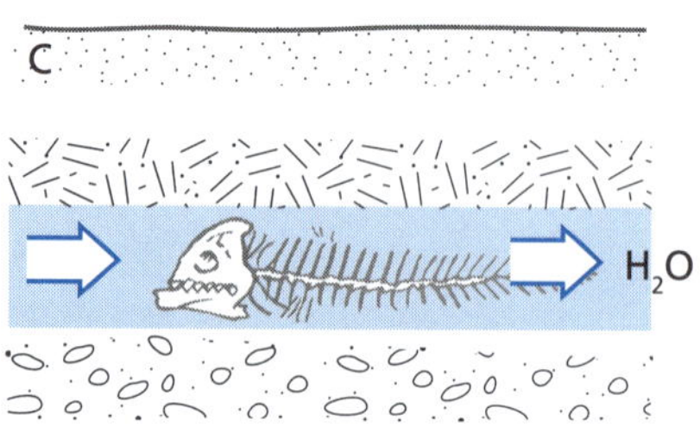

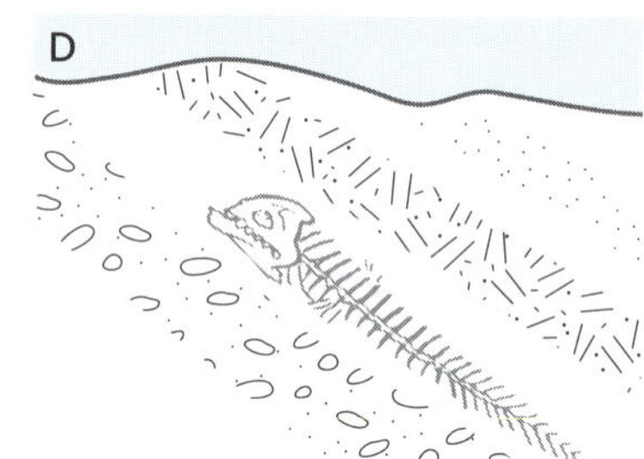

6 **a** Almost all living organisms can leave fossils, but usually only the hard parts of plants and animals fossilise. Explain. (2 marks)

b By far the most common fossil remains are those of shelled invertebrate sea-living creatures such as snails, corals and clams. Fossils of land animals are far scarcer. Explain. *Hint 3* (2 marks)

Hint 1: How were these sedimentary rocks formed?
Hint 2: This is the 'glue' which holds the different grains together.
Hint 3: What conditions are needed for fossils to occur?

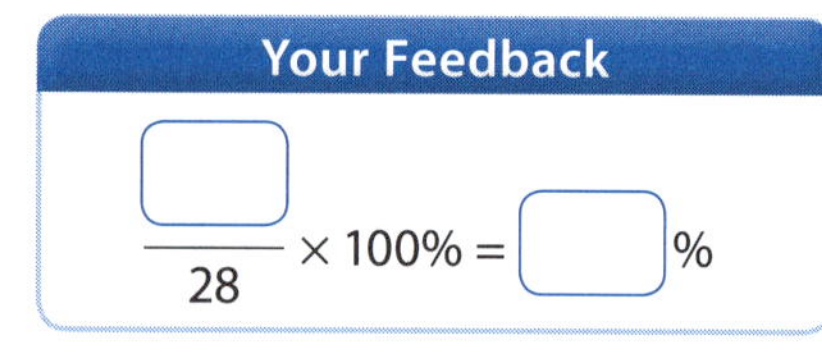

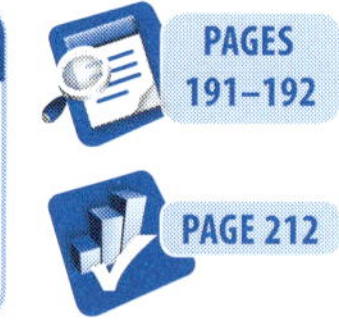
PAGES 191–192

PAGE 212

IGNEOUS AND METAMORPHIC ROCKS

Minerals and rocks

QUICK REVISION

1 Some rocks form as __________ magma cools, producing rocks containing crystals. These are known as igneous rocks. Some form __________ within the Earth's crust where temperatures are very __________, and could take thousands of years to cool down. As the magma slowly cools, __________ rock begins to form. This allows the crystals to grow much __________ Igneous rocks that are formed __________ the surface are called intrusive rocks. Granite is an intrusive igneous rock which shows large __________. Other igneous rocks form on the __________ of the Earth as lava cools down relatively __________, in hours or days. __________ is molten rock expelled by a volcano during an eruption; it can be considered to be magma that has had pressure removed from it allowing gases and dissolved liquids to __________. Those igneous rocks that cooled on the __________ are called extrusive rocks. The crystals in these rocks can be microscopically __________.
The type of rock obtained depends on what kind of lava the rock was formed from, and how __________ the rock cooled. A good example of an extrusive igneous rock is basalt. The small-grained crystals in basalt are just too tiny to see without a __________.

2 Metamorphic rocks are rocks that have __________ into other kinds of rock. These rocks may once have been __________ or sedimentary rocks. Under tremendous __________, and heat build-up, they slowly change. But they don't completely __________; the changes that affect these rocks occur in the solid state. If they did melt, then __________ rocks would result. They form deep within the Earth (12 to 16 km) where high __________ (100 to 800 °C), great pressure and chemical __________ act on them. The heat comes from magma and from heat released during chemical reactions within the Earth. The pressure comes from the many layers of __________ piled on top squeezing lower layers.

Metamorphic rocks often show __________, called foliation. This occurs when a rock is being shortened along one axis as it recrystallises. When a metamorphic rock is formed under pressure, its __________ become arranged in layers. Slate is a metamorphic rock formed from the __________ rock shale. Slate is useful for making roof tiles because its layers can be __________ split into separate flat sheets. Marble is another example of a __________ rock. It is formed from the sedimentary rock, limestone. Metamorphic rocks sometimes contain fossils if they were formed from a sedimentary rock and are not too altered, but the fossils are usually __________ out of shape.

3 The Earth's rocks do not remain the same forever. They are continually __________ because of processes such as __________ and large-scale earth movements. Over millions of years rocks are gradually recycled. This is known as the __________ cycle.

For example, sedimentary rocks can be changed into __________ rocks by heat and/or pressure. Later these rocks can be weathered and the pieces __________ away. These fragments may be deposited in lakes or __________, eventually forming new __________ rock. On the other hand, the sedimentary rock could be buried __________ within the Earth's surface where tremendous heat can __________ the rock altogether. Later, when this rock cools, an igneous rock forms. There are many routes around the rock cycle, some occurring faster than others.

Answers 1 liquid; deep; high; solid (igneous); larger; below; crystals; surface; quickly (rapidly); lava; escape (flee, evaporate); surface; small (tiny); fast (quickly, rapidly); microscope 2 changed; igneous; pressure; melt; igneous; temperatures; reactions; rock; layering; crystals; sedimentary; easily; metamorphic; squashed 3 changing; weathering (erosion); rock; metamorphic; transported (eroded); seas; sedimentary; deep; melt

IGNEOUS AND METAMORPHIC ROCKS

Minerals and rocks

REVISION SUMMARIES

1 **Igneous rocks** consist of one of the three main rock types, the others being sedimentary and metamorphic rocks. Igneous rocks form when **magma** or **lava** cools and solidifies. They may form either below the surface as intrusive (plutonic) rocks or on the surface as extrusive (volcanic) rocks. Magma is a high-temperature fluid and complex mixture, where the temperatures usually range from 700 to 1300 °C. Inside the Earth solid rock can melt to form magma depending on its temperature, pressure and composition. This magma can be derived from partial melts of pre-existing rocks in either a planet's mantle or its crust. There are over 700 types of igneous rocks, most having formed beneath the surface of Earth's crust.

Intrusive igneous rocks are formed when magma cools and solidifies within the crust of the Earth. Buried within pre-existing rock (called country rock), the magma cools slowly, resulting in rocks that are coarse-grained. The mineral grains in these rocks can generally be identified with the naked eye. **Extrusive** igneous rocks are formed on the surface of the crust. These rocks cool and solidify faster than intrusive igneous rocks, and are fine-grained.

Igneous rocks may be classified according to how they were formed, their texture, the minerals they contain and their chemical composition. This classification can provide important information about the conditions under which they formed.

The following table gives a simple classification of some igneous rocks according to their composition and where they formed.

		high silica content ⟵	⟶	low silica content
		felsic	intermediate	mafic
extrusive (very fast cooling)	glassy or very fine grained	obsidian/pumice	obsidian/pumice	obsidian/scoria
extrusive (fast cooling)	fine grained	rhyolite	andesite	basalt
intrusive (slow cooling)	medium grained			dolerite
intrusive (very slow cooling)	coarse grained	granite	diorite	gabbro

Obsidian is a naturally occurring volcanic glass formed as an extrusive igneous rock. It is formed when erupted lava from a volcano cools rapidly with minimum crystal growth. Pumice is formed when lava cools quickly above ground, trapping bubbles of gas. It is solidified frothy lava. This rock is so light that it floats on water. As this rock is so light, it is often used as a decorative landscape stone.

2 **Metamorphic** rocks occur when existing rocks are altered or transformed in a process called metamorphism ('change in form'). The original rock is subjected to **heat** and **pressure** (temperatures over 150 to 200 °C and pressures of 1500 times atmospheric pressure), causing significant physical and/or chemical changes.

There is a change in the particle size (**re-crystallisation**) of the rock during this process. This is due to high temperatures and pressures. High temperatures permit the atoms and ions in solid crystals to migrate and reorganise, while high pressure allows the crystals within the rock where they contact to intermingle (mix with each other). For example, in the sedimentary rock limestone the small calcite crystals change into larger crystals as it forms the metamorphic rock marble. In quartzite (metamorphosed sandstone), the often larger quartz crystals become interlocked. Foliation refers to the alignment of mineral grains in a metamorphic rock.

There are three types of metamorphism: contact, regional and dynamic.

- **Contact metamorphism** usually occurs around intrusive igneous rocks as the temperature increases when magma is injected into cooler surrounding country rock. Changes are greatest

where the magma comes into contact with the rock (temperatures are highest) and decrease with distance from it. Contact metamorphic rocks are usually called hornfels. These rocks do not often present signs of strong deformation and are usually fine grained.

- **Regional metamorphism** refers to changes in great masses of rock over a large area. Rocks can be metamorphosed simply by being at great depths below the Earth's surface, where there are high temperatures and tremendous pressures from the weight of overlying rock layers. Much of the lower continental crust is metamorphic. The extent of the deformation of these rocks may have destroyed any original features that could have identified the rock's previous history. Re-crystallisation of these rocks destroys the textures and any fossils that may have been present in the original sedimentary rocks.
- **Dynamic metamorphism** is a type of high-pressure/low-temperature change in rocks occurring along major fault zones of the Earth. Here pieces of crust are forced to slip past each other in opposite directions resulting in localised intense pressures on the rocks adjacent to the faults. The texture of the rock is changed but there is little or no change in mineral content.

The table gives some examples of metamorphic rocks and the parent rocks from which they were derived.

Original parent rock	Metamorphic rock formed
sandstone (s)	quartzite
limestone (s)	marble
shale (s)	slate
slate (m)	phyllite
phyllite (m)	schist
bituminous coal (s)	anthracite coal
granite (i)	gneiss

Type: s = sedimentary, m = metamorphic, i = igneous.

3 The **rock cycle** is an important idea in geology that describes the dynamic changes through geological time among the three main rock types: sedimentary, metamorphic and igneous. Each of the types of rocks is changed or reformed when changing conditions occur. An igneous rock such as basalt may break down and dissolve when exposed to wind and water or melt as it is pushed under a continent (subducted). Because of these driving forces, rocks do not remain in equilibrium and are forced to change as they encounter new environments. The figure shows the various stages of the rock cycle.

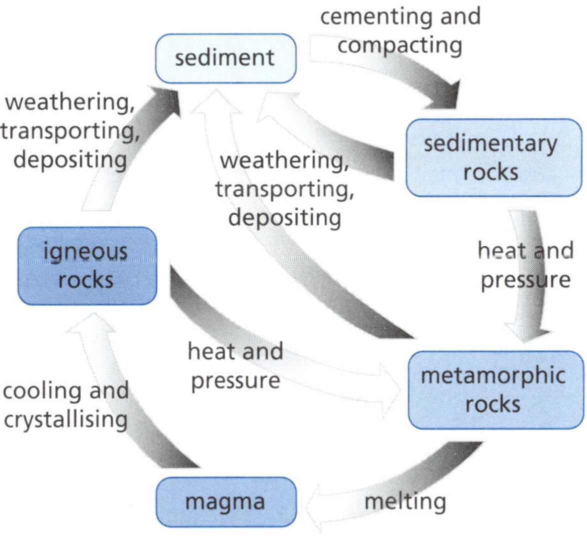

Checklist

Can you:

1. *Explain what igneous rocks are and identify the differences between intrusive and extrusive igneous rocks?* ☐
2. *Explain what metamorphic rocks are and explain the difference between contact and regional metamorphism?* ☐
3. *Describe the rock cycle and recognise that it is an ongoing process?* ☐

IGNEOUS AND METAMORPHIC ROCKS

Minerals and rocks

REVISION TEST

1 Complete the following statements.

a The three main types of rock are __________, igneous and metamorphic. (1 mark)

b Rocks buried deep underground that get heated and put under pressure are changed into __________ rock. (1 mark)

c If rocks underground get heated so much that they melt and turn into magma, then __________ rocks may form. (1 mark)

d Pressure forcing magma out of the ground through volcanoes can lead to solid rock, called __________ igneous rock. (1 mark)

e Magma cooling slowly underground forms solid rock called __________ igneous rock. (1 mark)

f The alignment of mineral grains in a metamorphic rock is called __________. (1 mark)

g The principal agents of metamorphism are heat and __________. (1 mark)

h Contact metamorphism results whenever __________ is in contact with other rocks. (1 mark)

i __________ metamorphism produces metamorphic rocks over large areas. (1 mark)

j Quartzite is a metamorphic rock that forms from the parent rock called __________. (1 mark)

k In the list limestone, basalt, gypsum and gneiss, the igneous rock is __________. (1 mark)

l A black, glass-like igneous rock which cools quickly preventing crystals forming is __________. (1 mark)

m An example of an igneous rock which floats on water is __________. (1 mark)

n The statue of William Charles Wentworth in the Great Hall at Sydney University is carved from marble. This is an example of a __________ rock. (1 mark)

o Sedimentary rock transforms to igneous rock by __________ and cooling. (1 mark)

p Metamorphic rock transforms to sediment by __________ and erosion. (1 mark)

2 Consider the following diagram.

original ground surface
X
present surface
country rock (original rocks)
igneous rock formed from molten magma
Y

a What process has caused the feature marked X? (1 mark)

b What type of rock is shown by Y? How was it formed? *Hint 1* (3 marks)

c Is the igneous rock harder or softer than the surrounding country rock? How do you know? (2 marks)

3 Complete the following table, using the words below, for this igneous rock chart. (6 marks)

andesite, basalt, diorite, gabbro, granite, rhyolite

	felsic	intermediate	mafic
extrusive			
intrusive			

IGNEOUS AND METAMORPHIC ROCKS

Minerals and rocks

REVISION TEST

4 Obsidian was valued in Stone Age cultures because it could be fractured to produce arrowheads. Even today, obsidian is still used for blades in surgery. What features allow obsidian to be used in this manner? *Hint 2* (2 marks)

5 **a** What is the rock cycle? (1 mark)

b The following diagram shows a simple rock cycle. Name the rock types labelled A, B and C. (3 marks)

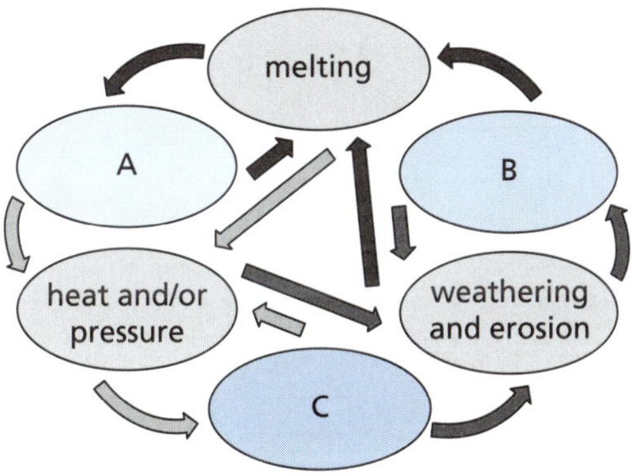

melting
A
B
heat and/or pressure
weathering and erosion
C

c A rock material went through being igneous then sedimentary. Is the next stage in the rock cycle for it to go through a metamorphic stage? Explain. *Hint 3* (2 marks)

6 The diagram shows a cross-section through a volcanic landscape. Describe the features of the igneous rocks formed at the positions labelled X, Y and Z. (6 marks)

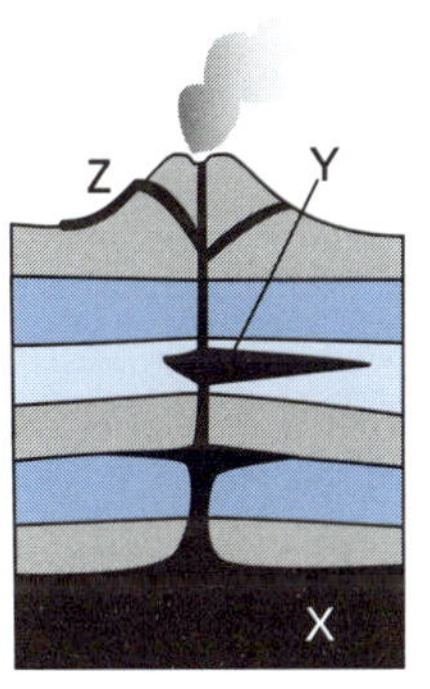

7 **a** What is the parent rock from which marble is derived? (1 mark)

b What are some features for which marble is used? (2 marks)

c Is marble always white? Explain. (2 marks)

d List two features that marble used for floor tiles or stair treads should have. (2 marks)

e The main mineral in marble is calcite (calcium carbonate). Suppose you were planning a kitchen benchtop made from marble. Why is it important that you regularly clean and polish it? *Hint 4* (2 marks)

Hint 1: Consider how this rock has changed as it came into contact with the igneous intrusion.
Hint 2: Consider the physical characteristics that put it in demand in this way.
Hint 3: Study the rock cycle. This will give you a clue.
Hint 4: Review your chemistry with the reactions of acids on carbonates.

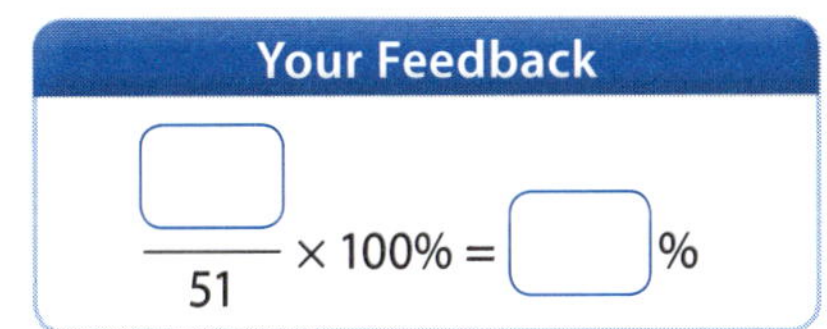

METALS AND ORES

Ores and environmental issues

QUICK REVISION

1 The earliest __________ that was mined by humans was gold. Gold is a __________ metal and does not react readily with other compounds present in nature. Gold is an example of a native metal. The following photo shows native gold in quartz.

Native metals are metals that can be found __________ in nature. Platinum and silver are also native metals. Iron can also be found free in nature but such iron is often alloyed with nickel and is usually of __________ origin. Metals such as calcium, sodium and aluminium are too __________ to exist as free elements. They are always found combined with other elements as minerals in the Earth's crust.

Minerals can be found in low concentration throughout the crust but sometimes they become concentrated and form an __________ body. If the concentration is high then it is called a high __________ ore. Mining companies can make large profits from high-grade ores. They continue to mine the ore body till a point is reached where any remaining ore is of such a __________ grade that continued mining becomes unprofitable and the mine is closed. Mines often reopen some time in the future when a new technology is developed which makes the mining __________ again. Alternatively a rise in __________ prices on the world stock markets can also make a low-grade deposit valuable.

2 Ore deposits form in a variety of ways over long periods of geological time. A good example of this is the lead and zinc ore deposit in Broken Hill in western NSW. It is estimated that this deposit __________ about 1.7 billion years ago when igneous activity led to the __________ of magma into the sediments followed by chemical precipitation of lead, silver and zinc sulfide minerals.

The following are some examples of the way ore bodies can form.

a Magmatic deposits. Molten rock is called __________ when it is underground and lava when it emerges onto the Earth's surface. Magma takes much longer to __________ and solidify underground than on the surface of the Earth. Large magma chambers underground gradually __________ and mineral crystals form and sink to the bottom of the magma chamber due to the __________ of gravity. Different minerals may mix together or form sequentially as the temperature drops. Such magmatic deposits form profitable __________ bodies. Chromite is an iron-chromium oxide mineral that deposits in magma chambers at __________ around 1500 °C. Cassiterite (tin oxide, SnO_2) is an example of a mineral that forms when the magma has cooled to between 500 and 800 °C.

b Hydrothermal deposits. These deposits are formed at great depths, near a __________ body of rock (e.g. cooling magma) and within the temperature range of 300 to 5000 °C. Hydrothermal solutions of __________ minerals are formed as water dissolves various minerals associated with the magma. These solutions __________ upwards (often in fractures in the rock layers) and minerals begin to crystallise and __________.

c Sedimentary deposits. Suspensions of insoluble materials gradually __________ under gravity in __________ bodies of water to form sediments. Ions of various metals such as iron and manganese underwent chemical __________ in the oceans of the Earth millions of years ago to form insoluble oxides. The metal oxides precipitated and formed __________ that became consolidated to form ore bodies. Banded iron formations are layers of sedimentary rock containing alternating layers of __________ oxides (such as haematite, Fe_2O_3 and magnetite, Fe_3O_4) and shales. These formed about 3.7 billion years ago when __________, produced by photosynthetic organisms in the oceans, reacted with iron ions and precipitated them.

d Placer deposits. Eroded material can be __________ by rivers to new locations such as a lake or a coastal basin. The __________ the river, the heavier the material it can transport. Small pieces of gold, for example, can be __________ from rocks and transported by a river to a new location. This alluvial gold may be deposited in places where the speed of the river __________ such as on bends in the river. Other lighter minerals can travel further and may be __________ in a lake or along the coast. Heavy mineral sands at some beaches are also __________ deposits. Examples include rutile (TiO_2), ilmenite ($FeTiO_3$) as well as gemstones such as zircons and garnets. The following photo shows some natural zircons.

Answers **1** metal; noble; free (uncombined); meteoric; reactive; ore; grade; low; profitable; metal **2** formed; intrusion; magma; cool; crystallise; pull; ore; temperatures; hot; dissolved; rise (move); precipitate; settle; quiet; reactions; sediments; iron; oxygen; transported; faster; eroded; decreases; deposited; placer

1 The discovery of **metals** marked a turning point in human civilisation. Gold metal was often found in quartz veins in rocks or as gold nuggets. A metal that can be found free in nature is called a **native metal**. Other native metals include silver, platinum, copper and iron. Native iron is usually found in meteorites from space. Most metals, including native metals, are present in the crust of Earth in compound form as minerals. For example, aluminium is too reactive a metal to exist in the native (uncombined) form. Instead it is found in the crust as aluminium oxide in the mineral bauxite.

Minerals may be present in small amounts throughout the crust. They may also be present in an **ore body**. An ore body contains a high mineral concentration which is profitable enough to mine and refine to extract the metal. Ore bodies can be classified as **high grade** or **low grade**. High-grade deposits contain a high percentage of the mineral and these deposits are very profitable. A low-grade ore may be currently uneconomic to mine but if the price of metals rise or extraction technologies become less costly, then these deposits can become profitable in the future. The following table gives examples of come common ore bodies and the metals that are extracted from them. The photo shows galena mineral from which lead can be obtained.

Ore	Metal extracted
bauxite	aluminium
chalcopyrite	copper
galena	lead

Ore	Metal extracted
haematite	iron
pitchblende	uranium
sphalerite	zinc

2 Ore bodies can form in many different ways over long periods of geological time. Ore bodies typically have more than one type of mineral. For example, the ore deposit at Broken Hill contains lead and zinc minerals.

Here are some examples of common geological processes that produce ore bodies.

- **Magmatic deposits.** Magma is molten rock inside the Earth's crust. If this magma does not flow onto the Earth's surface as lava, it will cool and crystallise underground. When specific minerals crystallise in large masses, a magmatic deposit will result. Chromite ($FeCr_2O_4$) is a mineral formed in magmatic deposits. The following diagram shows a magmatic deposit. Minerals crystallise out of cooling magma in a magma chamber.

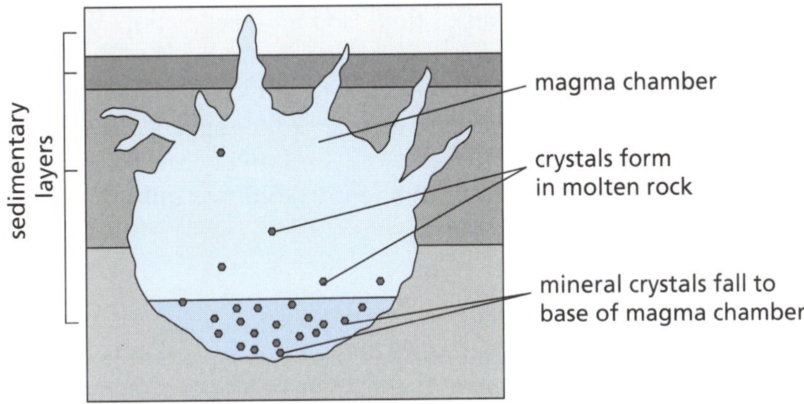

- **Hydrothermal deposits.** Groundwater may come near a hot body of rock. This water becomes very hot and it can dissolve minerals out of the rock. As the hot water moves into cooler areas of the crust, the dissolved compounds are precipitated forming crystals of various minerals. Chalcopyrite ($CuFeS_2$) and sphalerite (ZnS) form in this way.
- **Sedimentary deposits.** Many common ores were formed by chemical precipitation of minerals in shallow ocean basins. Most of these ore bodies contain oxides and hydroxides. Haematite (Fe_2O_3) and manganite ($MnO(OH)$) are examples of mineral deposits formed in this way. Many of these deposits show alternating bands of minerals and sedimentary rocks such as chert and shale.
- **Placer deposits.** Placer deposits are formed when minerals are transported by a river and deposited along the river or its bed, or in a quieter body of water such as a lake or at the river mouth. Alluvial gold (Au) and rutile (TiO_2) are common placer deposits.

Checklist ✓

Can you:

1 *Explain why ore bodies need to be high-grade deposits to make mining economical?* ☐

2 *Name four common ways in which ore bodies can form?* ☐

METALS AND ORES

Ores and environmental issues

REVISION TEST

1 Read the text below and answer the questions that follow.

> The discovery of metals marked a turning point in human civilisation. About 10 000 years ago in the Middle East, the first societies that used metals developed. The native or natural metals such as gold were the first to be used as they could be obtained from the ground without much further treatment. Gold was often found in quartz veins. Gold-bearing quartz was mined and crushed to release the gold grains. Gold nuggets could be readily hammered to produce ornaments or jewellery. Because of its relatively low melting point (1064 °C), gold could be melted, purified and cast into ornaments. Gold was quite soft and malleable and was usually only used as a decorative metal.

a Define the term *native metal*. (1 mark)

b Explain what is meant by the term *quartz vein*. *Hint 1* (1 mark)

c Identify three properties of gold that makes it suitable to create jewellery and ornaments. (3 marks)

2 Ore bodies often contain mixtures of different minerals. The properties of some selected minerals are shown in the table below.

Mineral	Chemical composition	Crystal form	Colour	Lustre	Hardness (Mohs scale)
pyrite	iron sulfide	cubic	bronze yellow	metallic, shiny	6–6.5
chalcopyrite	copper iron sulfide	cubic and wedge shaped	brassy yellow; iridescent	metallic	4–5
galena	lead sulfide	cubic	black–grey	shiny	2.5
sphalerite	zinc sulfide	cubic	black to brown	shiny	3.5–4
cassiterite	tin oxide	tetragonal	black-brown	glassy	6–7
chromite	iron chromium oxide	octahedral	black to brown-black	low sheen	5.5
molybdenite	molybdenum sulfide	flat, scaly	grey	metallic	1–1.5

Martin was given pairs of minerals from the table and asked to identify them. What distinguishing feature would allow Martin to identify the minerals in each of the following pairs?

a pyrite and galena (1 mark)

b cassiterite and chromite (1 mark)

c chalcopyrite and molybdenite (1 mark)

3 Name the metal that can be extracted from the following ore bodies.

a bauxite (1 mark)

b haematite (1 mark)

c pitchblende *Hint 2* (1 mark)

4 Examine the following diagram which shows rivers flowing from the mountains across a coastal plain and into the sea. Gold is shown in various deposits.

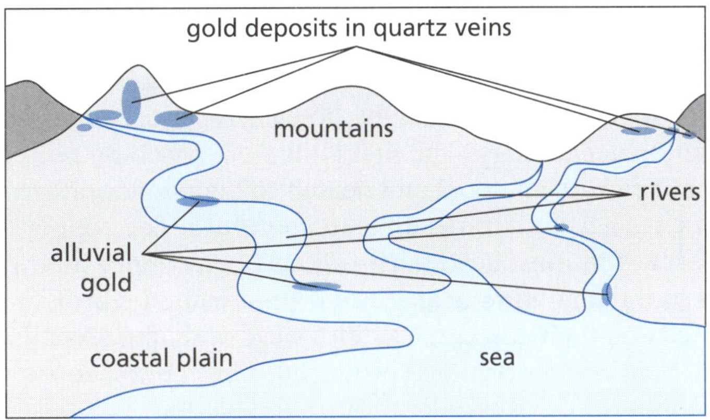

a Explain how the gold deposits in the quartz veins may have formed. (2 marks)
b Explain how the alluvial gold deposits have formed. *Hint 3* (2 marks)

5 The diagram below shows a close view of a rock retrieved from a mine site in Western Australia. The rock shows mineral bands between layers of shale.

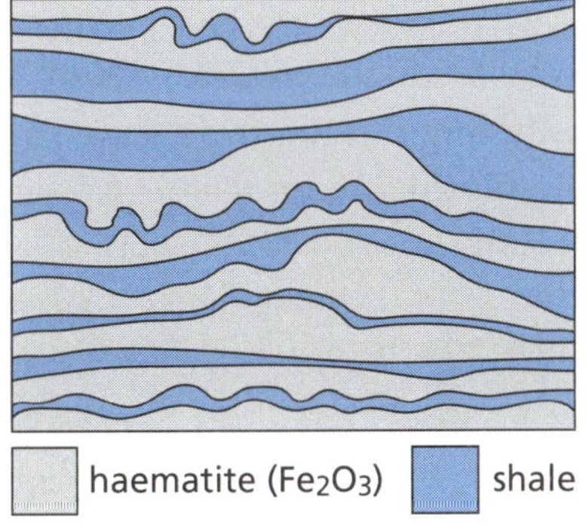

a Identify the type of ore deposit being mined at this site. (1 mark)
b Explain how such deposits form. (2 marks)
c What metal will be extracted from this deposit? (1 mark)

Hint 1: Quartz crystallises from the magma as the temperature of the solution drops.
Hint 2: This metal is radioactive.
Hint 3: Notice that rivers pass close to the quartz veins.

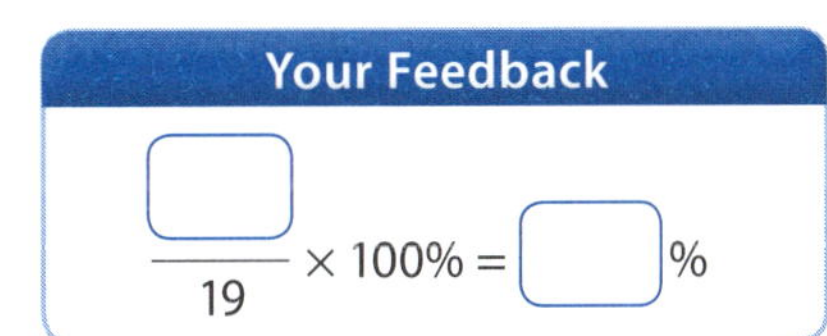

EXTRACTING METALS FROM COMMON ORES

Ores and environmental issues

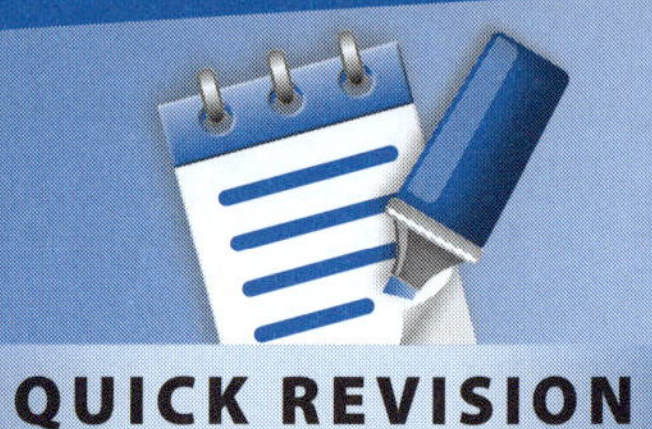

QUICK REVISION

1 Copper is a very important metal in our modern society. Copper has been known for about 8000 years. The earliest forms of copper were small amounts of __________ copper. These small amounts of native copper were fashioned into jewellery and fish hooks. People living in regions near modern Iraq also observed green and __________ minerals in the hillsides and discovered that when the rocks bearing these minerals were __________ in a wood/charcoal fire, small traces of metallic copper formed. Gradually these early observations led to the establishment of a roasting and smelting technology. The first kilns were based on pottery kilns. The __________ copper ores were mixed with charcoal and heated to high temperatures with the aid of bellows that provided extra __________ to raise the temperature. The copper minerals were converted into copper. The early kilns produced fairly __________ copper that was mixed with waste called slag. When cool the mixture of slag and copper could be broken up to release the copper which was re-melted and then __________ into various shapes. As kiln technology improved samples of __________ copper could be produced. This molten copper could be run out of the kiln and into clay __________ where it set into solid blocks.

Australia is the fourth-largest producer of copper in the world. Sites such as Mount Isa in Queensland and __________ Dam in South Australia are major areas of copper mining. Sites in New South Wales, Tasmania and Western Australia are also copper producers. Various copper ores are mined at these sites. Common copper minerals include __________ and chalcocite. The mined copper ore is first crushed and __________ into a powder. This process helps to remove the copper minerals from the unwanted rock. The mixture then undergoes __________ flotation to separate the minerals from the rocky __________ (see the diagram below). The minerals stick to the surface of bubbles in the froth flotation tank. These bubbles __________ to the surface and are scooped off and the minerals retrieved. The minerals are then converted to copper in a __________-temperature furnace. The molten copper is run into moulds and cooled back to __________ copper. The copper used in electrical wiring must be further __________ to eliminate any impurities. The impurities may represent about 2% of the mass. The impure copper is placed in an electrolysis cell and an electric current is passed through it. The electric current converts the dissolved copper to metallic copper which is 99.9% pure.

froth of air bubbles with minerals attached

collection of froth and mineral ores

bubbles rise to the surface carrying the tiny mineral grains

water

compressed air

Key: gangue (unwanted rocky waste) air bubble minerals attached to bubble

2 Aluminium is a widely used __________ in our modern society. Unlike gold and copper, aluminium was not discovered until recent times. The first samples of impure aluminium were made from aluminium compounds in 1825. Eventually __________ methods were developed

EXTRACTING METALS FROM COMMON ORES

Ores and environmental issues

QUICK REVISION

that could produce large quantities of aluminium fairly cheaply. Aluminium is now used in many products including guttering, cladding, window frames as well as kitchen appliances such as saucepans. Because aluminium is a low-__________ metal, it is commonly alloyed to produce lightweight yet __________ sheets for the hulls of yachts as well as aircraft bodies.

Aluminium is produced from __________ ore. Bauxite is an impure form of __________ or aluminium oxide. In Australia, large ore bodies are found near __________ in Cape York (northern Queensland) and in the Northern Territory at Gove. The ore bodies are easily mined as the mineral is found under a thin layer of topsoil. The ore body is red-brown in colour due to __________. The impurities are removed by grinding the ore to a fine powder and adding sodium __________ to dissolve the alumina. The impurities can then be __________ off and the dissolved alumina recovered by adding acid. The washed and dried alumina is a __________ powder. Alumina has a very high __________ point and so it is mixed with another compound called cryolite. This mixture melts at a much __________ temperature and so the costs involved are reduced. The aluminium is recovered from this molten mixture by passing electricity through carbon electrodes. The aluminium is __________ at these high temperatures and so can be run off into moulds to cool and solidify. The following photo shows the accumulation of bauxite ore after mining.

Answers **1** native; blue; heated; powdered; air; impure; cast; molten (melted); moulds; Olympic; chalcopyrite; ground; froth; gangue; rise; high; solid; purified **2** metal; electrolytic; density; strong; bauxite; alumina; Weipa; impurities; hydroxide; filtered; white; melting; lower; molten

EXTRACTING METALS FROM COMMON ORES

Ores and environmental issues

REVISION SUMMARIES

1 Copper minerals and metallic copper were discovered about 7000 to 8000 years ago in regions of Asia close to modern Iraq. The hills contained veins of native copper as well as green and blue copper ores. Between 7000 and 6000 years ago, people had discovered how to convert the copper ores into copper by a process called **smelting**. It is believed that this was done in pottery kilns where the powdered ores were mixed with **charcoal** (impure carbon) and heated to a high temperature by a blast of air. A molten slag formed which, on cooling, contained coppery grains. The cold slag was broken up to get the copper, which was then re-melted and **cast** in clay crucibles. Eventually, high-temperature closed kilns were developed to produce molten copper on a larger scale. Copper axe-heads, arrowheads and chisels were cast in clay moulds. Copper also began to be used to make plates, cups and cutlery.

Today Australia produces about 20% of the world's copper. Major mining sites are at Mount Isa in Queensland and Olympic Dam in South Australia where uranium, gold and silver are also mined. At Olympic Dam the ore body is mined using **underground mining** principles. The following diagram shows the steps in the mining process and the conversion of the copper ore to copper. Copper ore is mainly composed of chalcopyrite ($CuFeS_2$), bornite (Cu_5FeS_4) and chalcocite (Cu_2S) minerals. The copper minerals are separated from unwanted rock (**gangue**) by a process called **froth flotation**. The crushed ore is mixed with water and detergents and air is blown through the mix to produce a froth to which the copper minerals stick. The heavier gangue sinks to the bottom of the tank. Once the minerals are separated, they are heated to high temperatures in a furnace and reduced to copper metal. The copper produced from the furnace is further purified by a process called **electrolysis** involving the use of electricity.

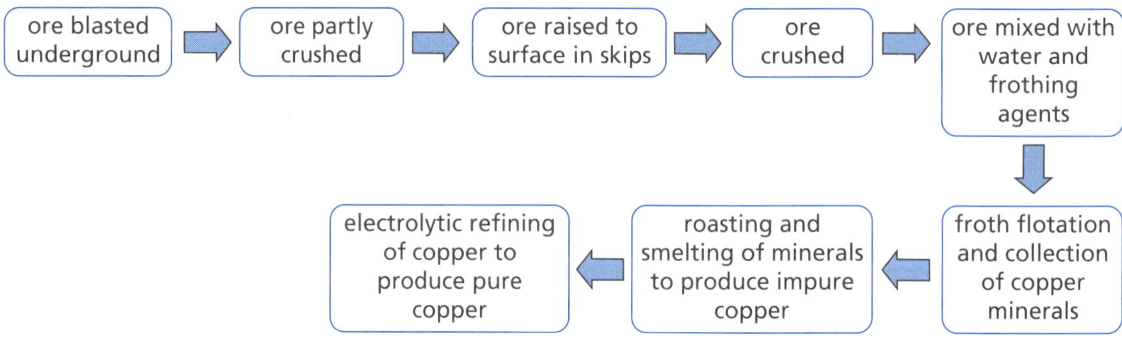

2 Aluminium is a very reactive, low-density metal and it does not exist as a native metal. It was discovered in 1825 when Hans Oersted produced an impure sample of the metal from aluminium chloride. Because of the difficulty of extracting aluminium from its ores, aluminium was initially more valuable than gold in the middle of the 19th century. An **electrolytic** method to make aluminium was developed in the 1880s. Today aluminium has many uses including saucepans, window frames, guttering, high-voltage electrical wires and in the construction of aeroplanes and yachts.

In Australia, **bauxite** ore is found at Weipa in northern Queensland as well as Gove in the Northern Territory and several other sites in southern Western Australia. Australia currently produces most of the bauxite in the world. Bauxite contains about 50% alumina (Al_2O_3). It is formed by the weathering of sedimentary rocks that contain a high proportion of aluminium minerals. Bauxite is usually found about 50 cm beneath the topsoil. When mined, bauxite resembles small red pebbles. The steps of conversion of bauxite to alumina are shown in the diagram on the next page. In order to extract the aluminium from the alumina, the alumina

must first be melted. This is achieved by mixing it with a substance called **cryolite** and then heating the mixture until it melts. The aluminium ions are converted to aluminium metal by passing an electric current through the molten liquid. The aluminium that forms is a liquid at these temperatures (~960 °C) and it sinks to the bottom of the steel furnace where it is run off into moulds.

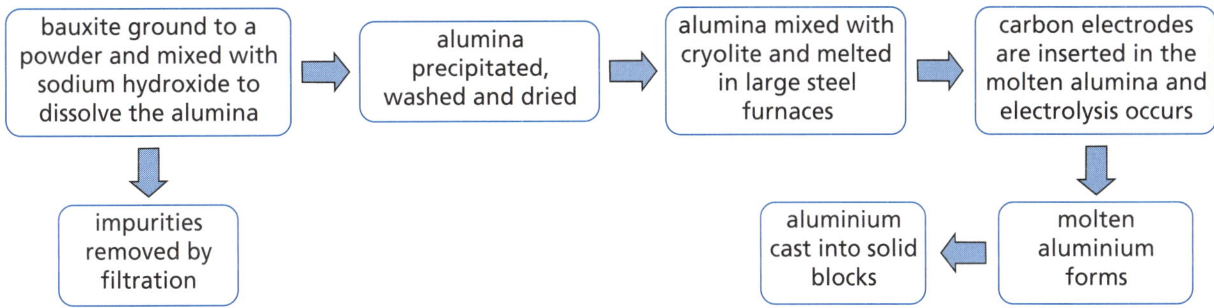

The following photo shows electrolytic refining of aluminium.

Checklist

Can you:

1. *Describe the stages of the production of copper from copper ores such as chalcopyrite?* ☐
2. *Describe the stages of the production of aluminium from bauxite ore?* ☐

EXTRACTING METALS FROM COMMON ORES

Ores and environmental issues

30 MINUTES

REVISION TEST

1 a Classify aluminium as a metal, semi-metal or non-metal *Hint 1* (1 mark)

b Name the ore from which aluminium can be extracted. (1 mark)

c Name the compound of aluminium that is present after the impurities in the ore are removed. (1 mark)

d Explain why large amounts of electricity are required in the extraction of aluminium from its ore. (1 mark)

2 a Classify copper as a metal, semi-metal or non-metal. (1 mark)

b Name two common ores from which copper can be extracted. (2 marks)

c Name the physical separation technique that is used to free copper minerals from the unwanted rock. (1 mark)

d Name a common use of pure copper in our society. (1 mark)

3 The following diagram shows the structure of an aluminium electrolytic furnace. It operates at temperatures around 960 °C.

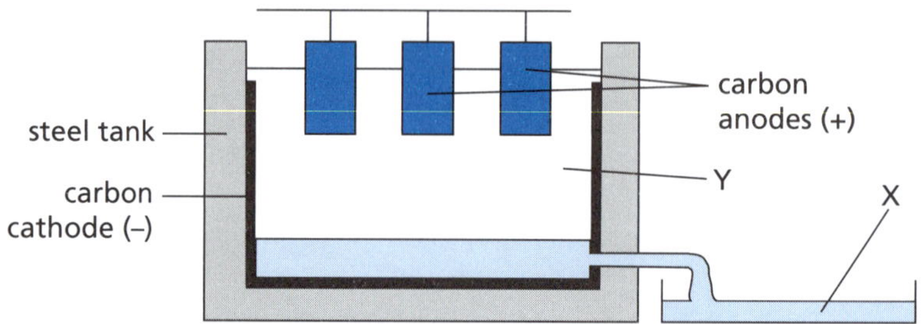

a X is a molten liquid which cools and solidifies. Identify X. (1 mark)

b Y is a mixture of two substances. Name them. (2 marks)

c Why must the temperature of the furnace be held at 960 °C? (2 marks)

d State two physical properties of X. *Hint 2* (2 marks)

4 Chalcopyrite is a mineral from which copper can be extracted. The chemical composition of pure chalcopyrite (by weight) is shown in the following data table.

% copper	% iron	% sulfur
34.6	30.4	35.0

a Calculate the theoretical weight of copper that could be extracted from 500 t of pure chalcopyrite. (2 marks)

b Explain why this mass of copper (calculated in part a) is never obtained in a normal extraction process. *Hint 3* (1 mark)

c Draw a pie (sector) graph of this numerical data. (3 marks)

5 A 1-t sample of bauxite is sent for processing to recover the alumina. Following the removal of impurities, the alumina recovered had a mass of 550 kg.

a Calculate the percentage by mass of alumina in the bauxite, assuming no alumina is lost in the processing. (2 marks) b Alumina is 79.4% aluminium by weight. Assuming no losses, calculate the maximum mass of aluminium that could be recovered from the 550 kg of alumina in part a. (2 marks)

6 True or false?

a Copper was first discovered by ancient people about 2000 years ago. (1 mark)

b Axe heads were made from copper before they were made from steel. (1 mark)

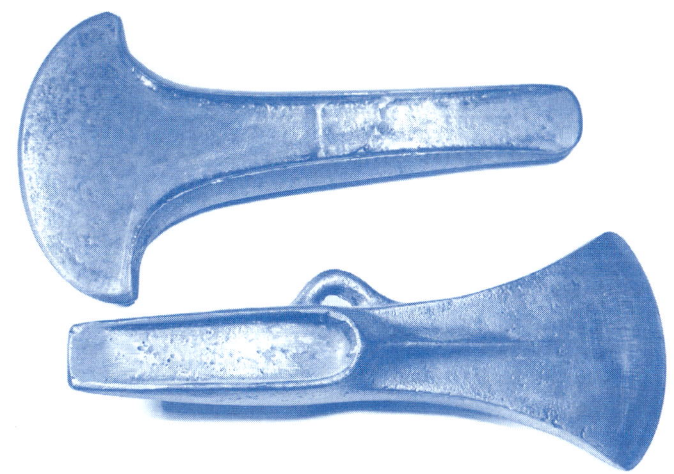

c Mt Isa is a copper ore mining site in Queensland. (1 mark)

d Chalcocite minerals are composed of the elements copper, iron and oxygen. (1 mark)

e Froth flotation is a useful technique for separating crushed ore from unwanted rock. (1 mark)

f Aluminium is found in nature as a native metal. (1 mark)

g Weipa is an important bauxite mining site in Australia. (1 mark)

h Bauxite ore bodies typically contain 50% by weight of alumina. (1 mark)

i Alumina is a mineral with the chemical formula AlO_2. (1 mark)

j Impurities are removed from bauxite by dissolving the alumina in sodium hydroxide and then filtering off the impurities. (1 mark)

Hint 1: Is aluminium shiny or dull?
Hint 2: Physical properties refer to its appearance, density, melting and boiling points and conductivity.
Hint 3: Think of all the steps involved in the extraction process.

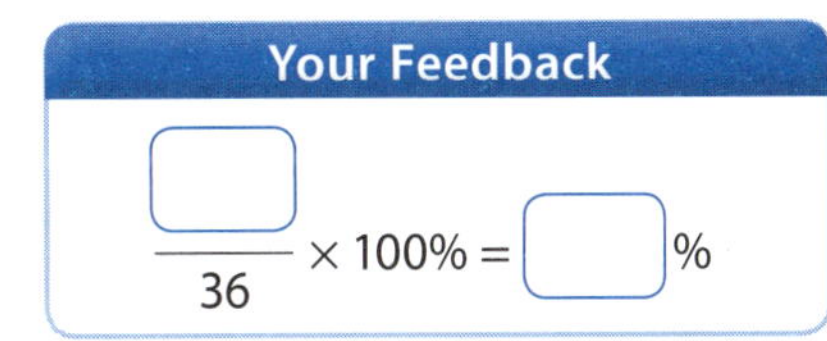

PAGES 194–195
PAGE 212

MAINTAINING OUR LOCAL ENVIRONMENT

Ores and environmental issues

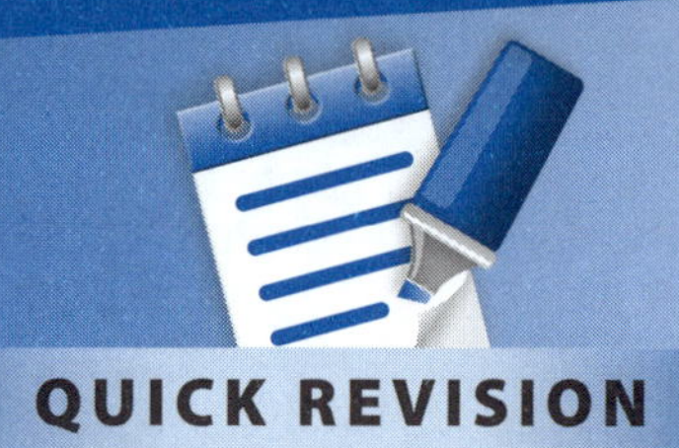

QUICK REVISION

1 The world's population has now reached over 7 __________ people. These people all have needs for food, __________ and clothing. In the developed world there is also a strong appetite for consumer goods, many of which will be __________ out once their usefulness has gone. The rate at which people are using natural resources cannot be sustained, unless we __________.

Recycling conserves __________ resources. For example, recycling a tonne of aluminium conserves up to 6 t of __________ ore and 14 megawatt-hours of __________. In other words, making a can from recycled aluminium rather than from bauxite saves 95% of the __________. As aluminium does not degrade or break down during the __________ process, it can be recycled over and over again. Not only will the raw materials last longer it is also __________ to recycle metals, such as aluminium, than to produce them from raw materials. Recycling helps reduce the consumption of water, energy, fuel and valuable natural resources, and cuts down on waste and __________ gas emissions. Councils and other organisations around the country have now provided special __________ for recyclable materials. At the sorting station, people and conveyor belts separate the plastic, glass, paper, aluminium and steel cans. These __________ are sent off to be recycled.

2 Until the importance of wetlands was discovered, many __________ that existed at the time of European settlement were destroyed. Then, many people thought it to be __________ land that could be filled in and turned to farming. A wetland is a land area that is __________ with water, either permanently or seasonally (wet for only part of the time and __________ periodically). It has characteristics that distinguish it from other types of ecosystems. Wetlands contain __________ that is adapted to its unique soil conditions and this includes aquatic plants. In wetlands the water may be freshwater, __________ or brackish. Wetlands include swamps, __________, billabongs, mangroves and bogs. They serve a number of functions in the environment, mainly water __________ (they filter pollutants and fertilisers), flood control and regulate river flow, they are spawning zones for some __________ of fish and they maintain shoreline stability. Wetland plants can provide increased __________ resistance of the soil mass in which they are planted and so reduce the wave energy levels reaching the shoreline. Wetlands are home to a wide range of plant and __________ life. As well, they are often beautiful places with cultural, historical and __________ values.

3 Pollution is introducing contaminants into the __________ environment that cause harmful changes. There are various types of pollution, such as chemical substances, or __________, such as noise, heat or light. Core samples from Greenland glaciers showed there was even pollution associated with Greek, Roman and Chinese metal production in __________ times, but at that time the pollution was far less and could be handled by __________. It was not until the __________ revolution that environmental pollution, as we know it today, had its beginnings. Large factories that __________ tremendous quantities of coal and other fossil fuels gave rise to unprecedented __________ pollution and to the large volume of industrial chemical discharges which added to the growing load of untreated human waste. Nature cannot handle these large-scale __________ in pollution. Faced with deteriorating air and water quality, world governments have passed laws making those who are responsible for the pollution also responsible for __________ it up.

Answers **1** billion; shelter (housing); thrown (discarded); recycle; natural; bauxite (aluminium); electricity; energy; recycling; cheaper; greenhouse (polluting); bins (receptacles, containers); materials **2** hectares (areas, square kilometres); wasted (useless, unproductive); saturated (covered); dry; vegetation (plants, flora); saltwater; marshes; purification; species (varieties); erosion; animal; educational **3** natural; energy; ancient; nature; industrial; burned (used); air; increases; cleaning

MAINTAINING OUR LOCAL ENVIRONMENT

Ores and environmental issues

REVISION SUMMARIES

1 Lead has been used for at least 5000 years. In ancient times it was a by-product of silver smelting and the Romans used it for water pipes, dishes, cosmetics, coins and paints. Winemakers in the Roman Empire would boil crushed grapes using lead pots or lead-lined copper kettles. Today, the major use of lead is for car lead batteries, which accounts for about 80% of consumption. Over 90% of the lead used in battery manufacture is **recycled** annually.

Many other metals can be recycled as well. The benefits of recycling these metals include the following.

- **Cost-effective.** Recycling metals is cheaper to produce than metal from ore extraction.
- **Conserves resources.** Ore reserves are not used up as fast, as there are other sources for the metal.
- **Waste reduction.** The amount of waste associated with primary extraction of the metal from the ore is reduced.
- **Human health benefits.** There is less opportunity for metals and associated garbage to end up in the waste stream, such as groundwater or atmosphere, where it could pose a health risk to people.
- **Reduces strain on landfill.** This helps conserve landfills for the garbage which can't be recycled.

Steps have been taken by water authorities around the country to recycle water. Recycled water can be used for a variety of non-drinking water purposes such as flushing toilets, watering gardens and parks, washing cars and for other outdoor uses. Some industries, such as steel producers, use a tremendous amount of water. Recycled water puts less strain on good, quality drinking water.

Continual improvements in recycling technology means more materials can be efficiently recycled than ever before. This includes paper, glass, plastics, computers and mobile phones. This is environmentally responsible and helps support the environment.

2 In Australia, the primary responsibility for **managing wetlands** and the animals and plants in them rests with the appropriate landholders or land managers. Individual state and territory

(cont.)

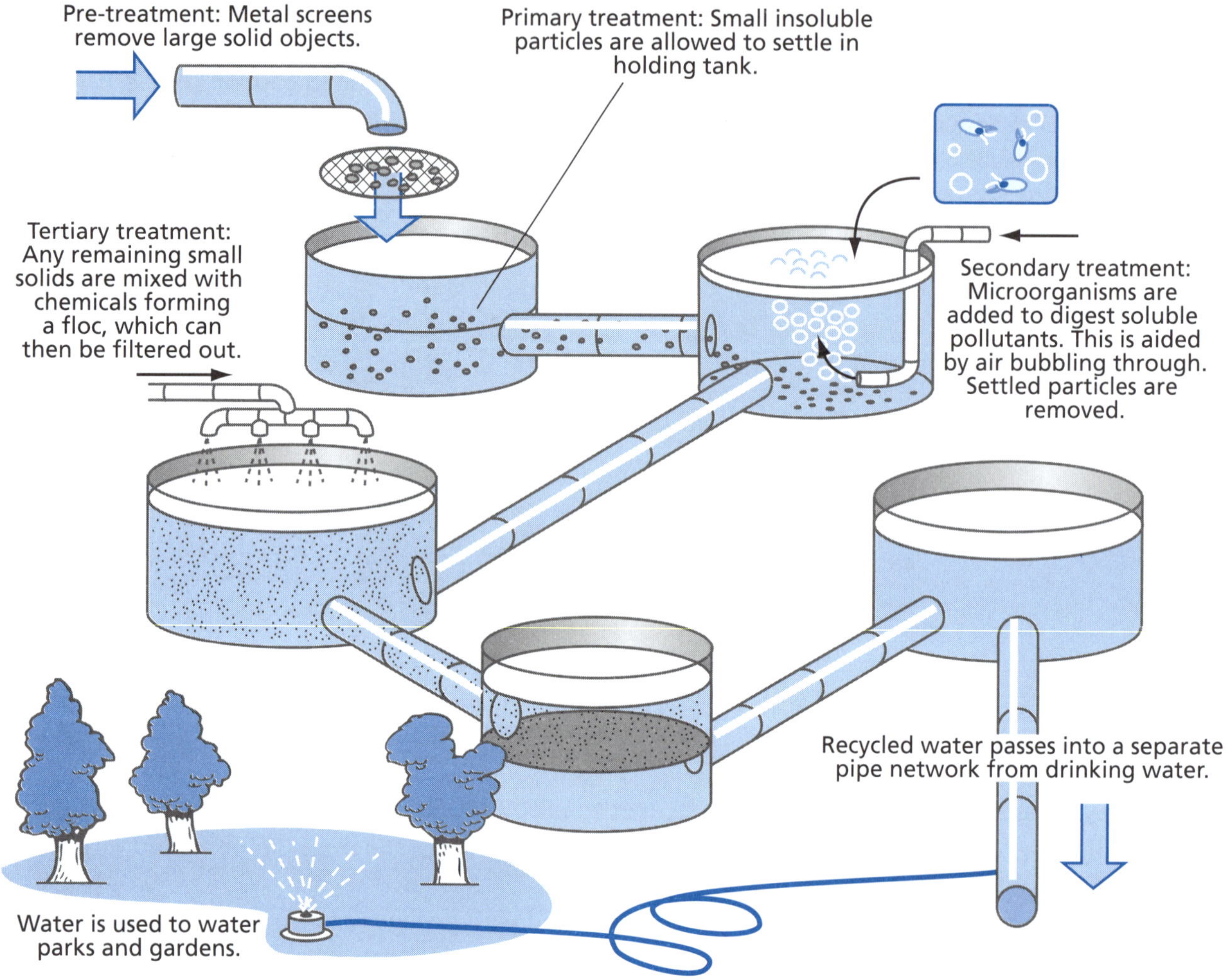

MAINTAINING OUR LOCAL ENVIRONMENT *(continued)*

Pre-treatment: Metal screens remove large solid objects.

Primary treatment: Small insoluble particles are allowed to settle in holding tank.

Secondary treatment: Microorganisms are added to digest soluble pollutants. This is aided by air bubbling through. Settled particles are removed.

Tertiary treatment: Any remaining small solids are mixed with chemicals forming a floc, which can then be filtered out.

Recycled water passes into a separate pipe network from drinking water.

Water is used to water parks and gardens.

governments have the legal responsibility for natural resource management and to check that these wetlands are managed properly.

Wetlands are important in many ways.

- They are very productive **ecosystems**. They are able to capture energy and provide food for many animals.
- They are breeding grounds for wildlife, particularly waterbirds, and nursery areas for fish.
- They provide important refuges for wildlife, especially in times of drought.
- They support a wide variety of plants and animals, forming different habitats and ecosystems.
- Wetlands provide vital **habitats** for some species of threatened animals. Some species of animals and plants are only found in wetlands.
- They provide a natural water balance in the landscape and help to provide protection against floods.
- By **filtering pollutants** such as sediments, nutrients, organic and inorganic matter and bacteria, they are providing water quality protection in the catchment area.
- Many migratory waterbirds that breed in the northern hemisphere find refuge in Australian wetlands. Thousands of these birds inhabit Australian wetlands every year.

- Aboriginal people identify a cultural significance with many wetlands.
- Wetlands are naturally beautiful places, providing opportunities for recreational activities such as swimming, boating, bird watching and bushwalking.

In addition, they provide opportunities for scientific research allowing us to find out more about their importance and the role they play in the ecosystem.

3 A **pollutant** is a waste material that pollutes air, water or soil. Three factors determine the severity of a pollutant: what it is made from (its chemical nature), how much there is (its concentration) and how long it remains around in the environment (its persistence).

Natural disasters, such as floods, tsunamis and earthquakes, can result in large-scale pollution. However, especially with technological advances in recent times, people have increasingly been responsible for wide-scale pollution as well. The massive earthquake and subsequent tsunami that hit Japan in 2011 had a devastating effect. To make matters worse, this terrifying natural disaster sparked a human-caused crisis, as radiation leaked from crippled reactors at the local nuclear power plant. Even though Japan had taken steps in readiness for any potential catastrophe, it nevertheless set off one of the worst nuclear disasters in history.

But pollution doesn't just accompany disasters. At the local level wastes discarded onto streets and parks can find their way into stormwater drains, killing fish and other animals many kilometres away. One of these wastes is plastic bags and ties, which take thousands of years to break down and can choke sea birds, turtles and other animals. It is therefore important that we dispose of wastes appropriately and not just discard them for others to clean up after us. The following photo shows plastic bottle waste washed up on the beach.

The borderless nature of atmosphere and oceans inevitably results in pollution in one area being caused by the behaviour of people in another area. Pollutants have been detected in various ecological habitats far removed from industrial activity such as the Arctic and Antarctic.

Checklist

Can you:

1 *List important reasons for recycling mineral/metal resources including water?* ☐

2 *Identify the importance of wetlands and why they should be maintained?* ☐

3 *Recognise the importance of preventing pollution?* ☐

MAINTAINING OUR LOCAL ENVIRONMENT

Ores and environmental issues

REVISION TEST

1 Glass recycling is both simple and beneficial. Give at least five reasons for recycling glass bottles. *Hint 1* (5 marks)

2 Wetlands are a critical part of our natural environment. Give at least four reasons for preserving wetlands. (4 marks)

3 Suggest how wetlands may: *Hint 2*

a control erosion (2 marks)

b prevent flooding. (2 marks)

4 The following column graph shows the paper consumption for an advanced country.

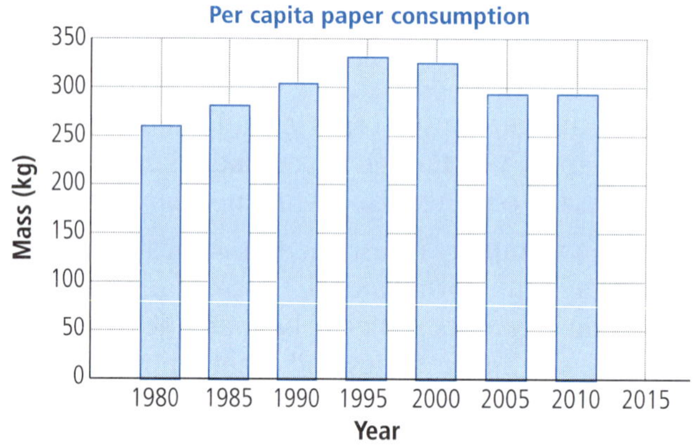

a What is meant by the term *per capita*? (1 mark)

b How much paper was used by each person in 1990? (1 mark)

c Does this mean each person used this amount of paper? Explain. (1 mark)

d In a city of 3 million people, how much paper was used in 2010? Answer in tonnes. (2 marks)

e Estimate the per capita paper consumption for 2015. (1 mark)

f Only about 10% of the paper consumed in a year is preserved in the form of filed documents, photographs, books and magazines. The rest finds its way into waste. How much paper per person was discarded in 1995? *Hint 3* (1 mark)

g It was said that the widespread use of computers would lead to a decrease in the use of paper. Is there any evidence for this from the graph? (1 mark)

5 Given that paper grows on trees, does it matter that we consume more paper than many other nations or that we don't recycle as much as we could? (2 marks)

6 Why might nuclear accidents and oils spills have long-term consequences? *Hint 4* (4 marks)

7 Since 1950, plastics have been a constant presence in our daily lives. They are found everywhere and globally we use more than 260 million tonnes of plastic every year. Describe two problems that pollution by plastics can cause. (2 marks)

8 The bar graph shows the waste composition of a typical garbage dump.

a What percentage of a typical landfill consists of metal? (1 mark)

b The bar for garden waste has been left off this graph. What is its value? (1 mark)

c Comment on how this graph might look if all recyclable material was recycled. (3 marks)

9

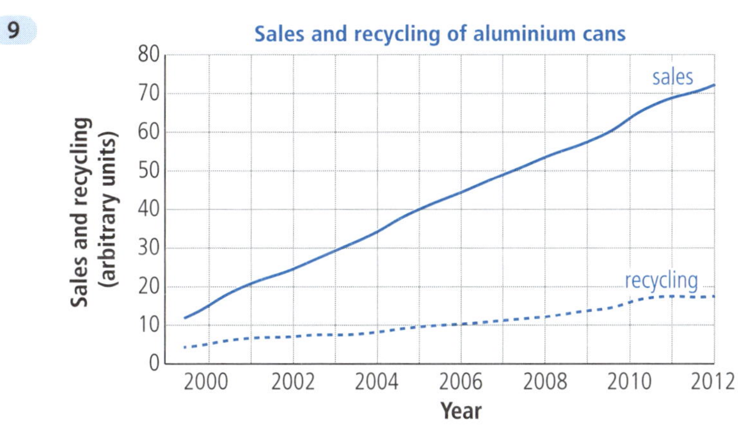

The line graph below shows sales and recycling of aluminium cans in an advanced country.

a Describe what the graph is showing. (2 marks)

b Why is it important that more aluminium cans be recycled? (3 marks)

c Describe two possible ways to increase the recycling rate for this country. (2 marks)

10 The table gives the distribution of Australia's wetlands. (Only those states/territories having over 1 million ha are listed.)

a Draw a column graph showing this information. (4 marks)

b Would it be true to argue that the differences between the wetland areas in each state/territory that are deemed nationally important and worth protecting depend on the political views of the various governments? Explain. *Hint 5* (3 marks)

State/territory	Area (millions of ha)
Queensland	42.9
South Australia	4.2
Northern Territory	4.0
Western Australia	2.6
New South Wales	2.3

Hint 1: Many of these factors are also true for recycling other substances, such as metals and paper.
Hint 2: Think about the types of plants growing in these wetlands and the amount of water they do and can hold.
Hint 3: If only 10% is kept, what percentage is thrown away?
Hint 4: Consider what you may read in newspapers about nuclear accidents and oil spills.
Hint 5: Consider the size of the state or territory and where it is located.

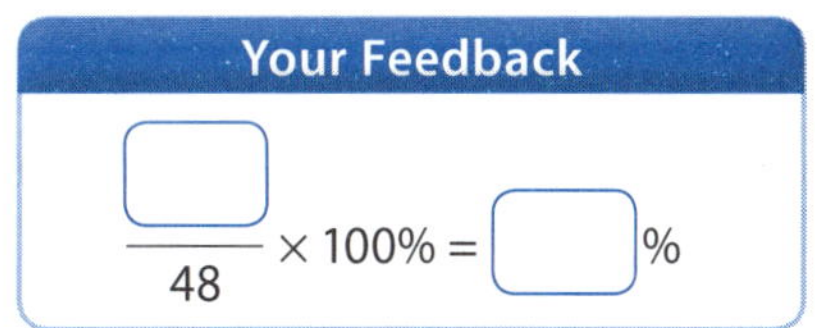

SUSTAINABILITY

Ores and environmental issues

QUICK REVISION

1 Agriculture covers farming (both __________ and animals), grazing and tending orchards, vineyards and timberland to produce food, feed, fibre and other goods by the systematic raising of __________ plants and animals.

In the first few decades after European settlement in Australia, farms developed around the early __________. Farmers bred sheep and mainly grew wheat crops that had originally been brought from __________. With road and rail developing, new tracts of land were __________ and farmers and squatters slowly moved inland. The Australian dry climate and infertile __________ presented challenges to farmers from the beginning. This includes water availability and __________ management. __________, such as superphosphate and nitrates, are still widely needed to improve soils. Farmers also have to deal with soil __________ and salinity. While wool and wheat were dominant in the __________ by the 20th century there was a greater diversity into beef and __________ cattle, and a wide range of grain, fruit and __________ crops. This included __________ cane in Queensland and fruit and wine __________ in the Riverina. The following photo shows the harvesting of a sugar cane crop.

Gradually, farmers learned that different areas suited different __________ of farming. Sheep and cattle grazing now takes up most of Australia's __________ land. Sheep are mostly found in New South Wales, Western Australia and Victoria. Queensland and New South Wales are Australia's main __________ cattle producers, with a significant contribution from the Northern Territory. Most __________ cattle farming is located in the southern states, especially Victoria.

Science and technology has made great improvements to farming in Australia, making it a world leader in __________ and efficiency. This has been helped with scientific advances in fertilisation, genetics, irrigation and disease control. __________-resistant crop strains have been developed and animals bred that are heat and insect tolerant while still producing quality meat, milk, eggs or wool.

2 Selective breeding is the process of systematically __________ plants and animals for particular traits and __________. Such breeding aims to establish and maintain __________ traits that plants and animals will pass to the next generation. In other words, the breeder tries to combine animals or plants in such a way that the offspring show better characteristics than their

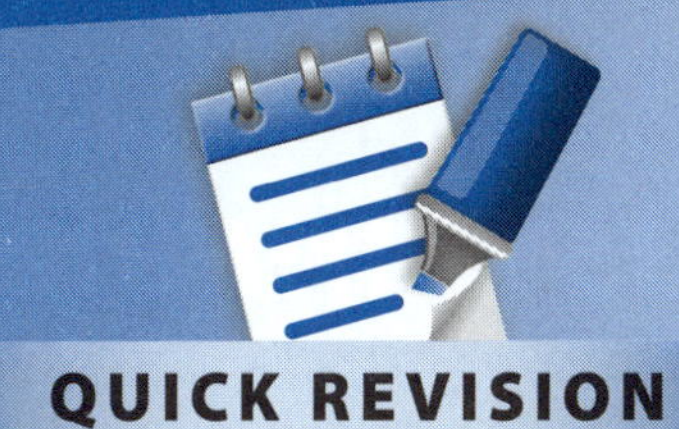

SUSTAINABILITY

Ores and environmental issues

QUICK REVISION

__________. Farmers have been selectively breeding for __________ of years. Dairy farmers breed cows that produce better quality and greater quantities of __________. Beef cattle producers want animals that put on a lot of __________ in a short time period. Horses bred for racing need to be __________ runners. Arab horses (as shown in the following photo) have been selectively bred for their racing abilities.

With the __________ breeding of pets and show animals, breeders try to make the animals resemble the written standards given for the breed. This may include their colour or markings, their temperament and their health.

3 Before Aboriginal people arrived in Australia, fires were haphazard and often started by __________ strikes. Over time, many __________ altered resulting in plants and animals either adapting to these fires or gradually dying out from the region. Aboriginal people changed the natural pattern and timing of these fires. They lit fires regularly and this changed the landscape in many parts of the __________. They used fire to:

- __________ prey, making them easier to catch
- encourage the __________ of fresh leaves, shoots and grass, attracting animals that could be easily __________. These __________-heat bushfires created an open scrub ecosystem. Plants that survived __________ to these conditions. Many native plants developed specialised survival characteristics while some need fire to regenerate or to release their __________. These fires lessened the frequency of producing more intense heat which would destroy even __________-tolerant plants.

Answers **1** plants; domesticated (tamed); towns (settlements); Europe; cleared; soil; drought; fertilisers; erosion; 1800s (19th century); dairy; vegetable; sugar; growing (production); types (kinds); agricultural; beef; dairy; productivity; drought **2** breeding; characteristics (features); favourable (useful, advantageous); parents; thousands; milk; muscle (meat); faster; selective **3** lightning; ecosystems (environments); country (nation); trap; growth; hunted (caught; killed); low; adapted; seeds; fire

SUSTAINABILITY

Ores and environmental issues

REVISION SUMMARIES

1 In the past and currently in many poor regions of the world, when people didn't know much about **sustainable** farming techniques they farmed the land until there were very few nutrients left. Combined with the lack of vegetation holding the soil together, especially trees, wind and water **erosion** would remove the valuable topsoil rendering the land useless for further agriculture. Intensive farming in fragile environments has taken its toll on natural resources. This has led to greater awareness of the need to use agricultural land sustainably and to maximise yields without compromising the health and productivity of the soil.

Sustainable agriculture is the practice of farming using **scientific principles**, especially the relationships between organisms and their environment. This is an integrated system of plant and animal production practices that will last in the long term. Agriculture has changed dramatically, especially over the last half century. Food and fibre production has soared, partly due to new technologies, mechanisation and increased chemical use. As well, fewer farmers with reduced labour demands can now produce these agricultural products more efficiently. Sustainability relies on the principle of meeting the needs of the present without compromising the ability of future generations to meet their own needs.

A number of factors need to be considered when trying to farm sustainably.

- Farmers need to avoid **irreversible changes** to the land, such as erosion. No more nutrients and resources should be taken from the environment than can be replenished. This includes not using more water than can be replaced regularly by rainfall or river flow. This will ensure that a single area will be able to produce food indefinitely.
- Consider the three Rs: **reduce, reuse, recycle**. Make the most efficient use of non-renewable resources and on-farm resources. Where appropriate, integrate natural biological controls and cycles.
- **Diversify** by raising both livestock and crops, and taking advantage of a mutually beneficial relationship between them. For instance, manure from livestock can fertilise crops; in return, use some crops to feed the livestock.
- Use varieties and breeds of animals and plants that are well-adapted to the local conditions.
- Use **companion planting** and green manures to keep the land perpetually fertile and to prevent topsoil loss. Companion plants are plants that together assist each other's growth in some way. For example, some plants can keep away pests, can attract beneficial insects that keep predators down, can increase nutrient uptake or can encourage pollination. A green manure is basically a crop that can be ploughed back into the ground while still green. The nitrogen-fixing bacteria found in the root nodules of the legume crop will help break down the plants, improving soil structure.
- Rotate crops and pasture. **Crop rotation** means growing a series of dissimilar types of crops in the same area in adjacent seasons. This reduces the build-up of diseases and pests that often occurs when only one type of plant is continuously grown. Nitrogen in the soil can be replenished by using green manure in sequence with cereals and other crops.

2 Artificial selection (or **selective breeding**) is the intentional breeding for certain traits (desirable characteristics or features), or a combination of traits. Humans have been selectively breeding animals and plants for millennia. There is no real genetic difference between artificial and natural selection. The selection process is termed *artificial* when humans intentionally influence the process.

Brassica oleracea (wild cabbage) is native to coastal southern and western Europe. Over several thousand years it has been selectively bred into a wide range of cultivars (as shown below) and

is now an important human food crop plant. Bred animals are known as breeds, while bred plants are known as varieties or cultivars. Crossing closely related animal species results in a crossbreed, while crossbred plants are called hybrids.

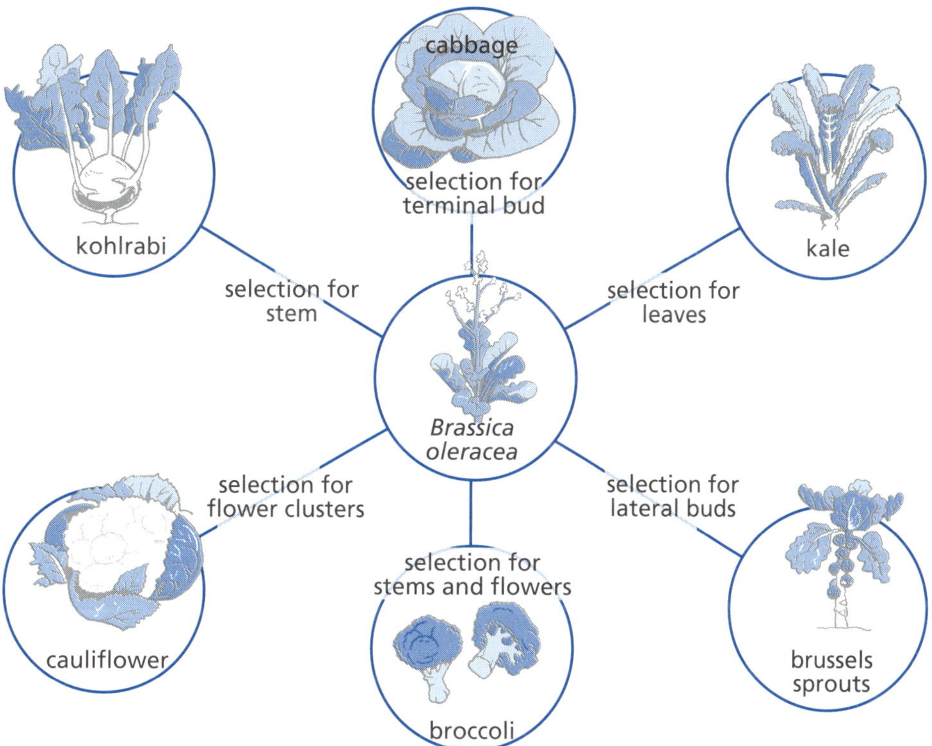

3 Australia is a dry continent and fires are a regular feature of the landscape. Aboriginal peoples regularly lit low-intensity fires to catch or attract animals, or for ceremonial purposes.

Plants respond differently when burned by a **bushfire**. Some die, others sprout new shoots, while others still appear not to have been affected. What occurs depends on the species of plant, on the intensity of the fire or on other environmental conditions before and after the fire.

Some species (e.g. *Xanthorrhoea*) flower copiously soon after fire, while others (such as some *Eucalyptus* species) can take years to make their first seeds. Other plants can **regenerate** from germinating seeds buried in the ground, stimulated by heat, smoke or the reduced competition following fire. Some woody plants of heaths and forests are killed by low-intensity fires; however, many bear canopy-borne cones or capsules that can open and liberate their seeds after a blaze. Examples of these plants include *Eucalyptus regnans* in Victoria and Tasmania, *Banksia ericifolia* around Sydney, *Banksia ornata* in South Australia and *Callitris* species in various parts of the nation.

Checklist
Can you:

1 *Identify the changing agricultural practices that have occurred in Australia?* ☐

2 *Define 'selective breeding' and explain why it is used?* ☐

3 *Recognise that fire is a regular feature in the Australian ecosystem, and that many plant species have adapted to it?* ☐

SUSTAINABILITY

Ores and environmental issues

REVISION TEST

1 a Explain the principle of crop rotation. (1 mark)

b Give two advantages for crop rotation. (2 marks)

c Legumes are plants that have bacteria in root nodules which can turn atmospheric nitrogen into a form plants can use. Explain how legumes can improve the soil by including them in a crop rotation program. *Hint 1* (2 marks)

2 In the late 1950s in Australia, the use of dual-purpose chickens for both egg and meat production ceased. Growers then started developing new poultry strains specifically for meat production, and others for egg production.

a Give a feature that would be important for chickens bred for meat. (1 mark)

b Give a feature that would be important for chickens bred for eggs. (1 mark)

3 Advances in abalone domestication and production techniques have risen sharply in Australia in the last decade. A challenge in abalone farming is to commercially create a hybrid between the two native species: greenlip abalone (*Haliotis laevigata*) and blacklip abalone (*Haliotis rubra*). These hybrid crosses are being farmed in southern Australia. Suggest a reason for farmers choosing to farm a hybrid animal. *Hint 2* (2 marks)

4 Explain how the following diagram shows selective breeding on a fish farm. (2 marks)

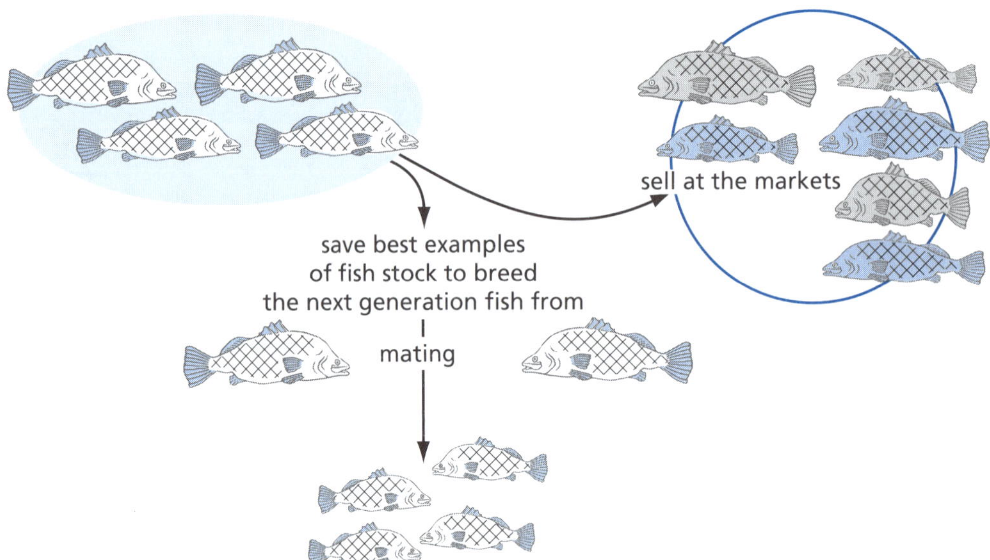

5 Suppose you are a plant farmer. Suggest five traits you would try to incorporate into your crop plants by selective breeding. *Hint 3* (5 marks)

6 Farmers often breed for the following features:

- growth rate
- survival rate
- quality of plant/animal matter
- being able to produce young in great numbers.

Give one reason for each to show why these features are important. (4 marks)

7 A corn farmer drew the following graph. Describe what the farmer is trying to do using this graph. (2 marks)

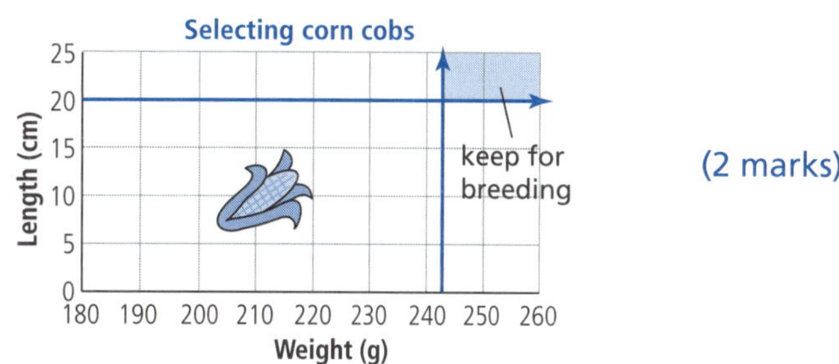

8 Describe two advantages a fire can have in a region for animals. (2 marks)

9 Many Australian native plants have adapted to survive low-intensity fires. Comment on each of the following adaptations and why it may be an advantage to the plant.

- **a** After fire, the Mountain Grey Gum tree (*Eucalyptus cypellocarpa*) produces a mass of shoots of leaves from burnt trunks and boughs, making it look like a black stick completely covered with young, green leaves. (2 marks)
- **b** *Acacia myrtifolia* seeds have a hard, shiny coating. These seeds need fire to germinate. (2 marks)
- **c** Grass trees (*Xanthorrhoea* species) are slow growing. While their grass-like leaf crowns burn readily, their tightly packed leaf bases will not. Thus the stem and apex of most mature plants survive a fire and quickly regenerate the crown. (2 marks)

10 A fish farmer was selectively breeding fish. The results are shown in the following graphs. Describe what the graphs are showing and what the farmer is trying to achieve. (4 marks)

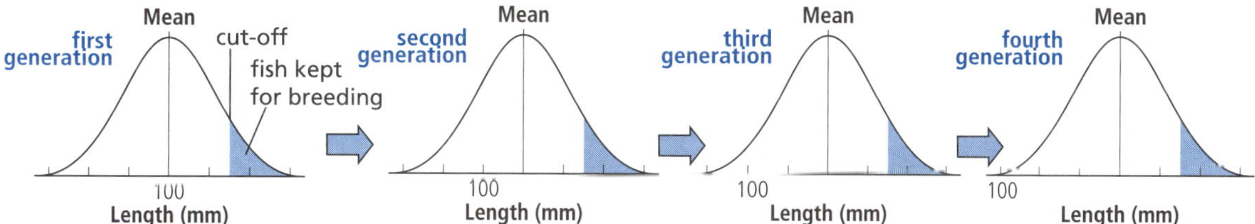

11 What feature does the following plant have to survive a fire? *Hint 4* (2 marks)

new shoot

ground level

tuber

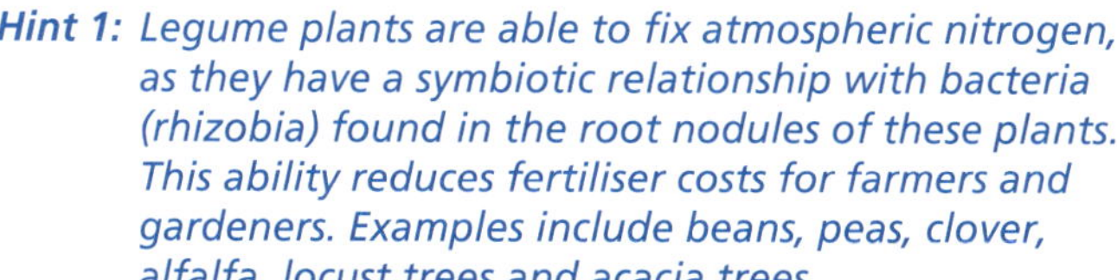

Hint 1: Legume plants are able to fix atmospheric nitrogen, as they have a symbiotic relationship with bacteria (rhizobia) found in the root nodules of these plants. This ability reduces fertiliser costs for farmers and gardeners. Examples include beans, peas, clover, alfalfa, locust trees and acacia trees.

Hint 2: What feature would a hybrid animal have that individually the two separate species would not have altogether?

Hint 3: Assume you were a plant farmer. What favourable features would you want your plants to have?

Hint 4: Soil can be a good insulator, preventing heat from progressing too far underground.

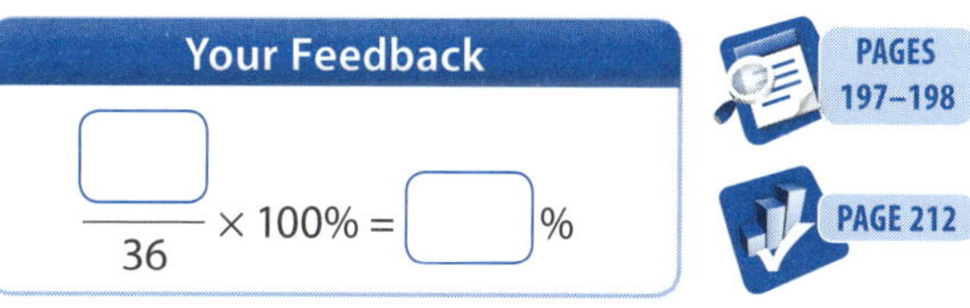

PAGES 197–198

PAGE 212

KINETIC AND POTENTIAL ENERGY

Energy

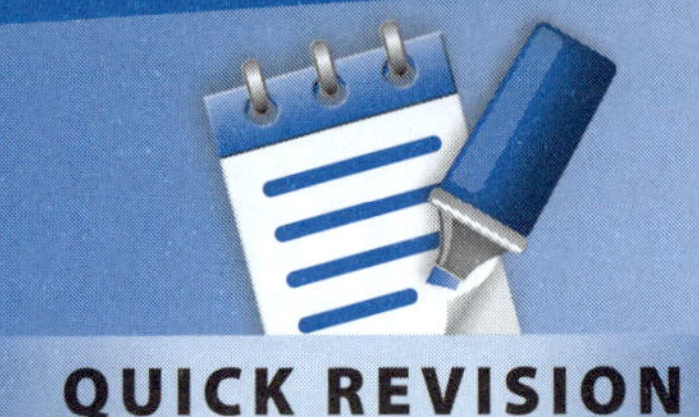

QUICK REVISION

1 An object has __________ energy when it is moving. Wind possesses kinetic energy as the air __________ are moving. The faster they move, the more kinetic energy they have. Wind turbines (see photo) are powered by the kinetic energy of the wind.

Very fast winds in a cyclone cause immense __________ because they have very high kinetic energy. Sound waves carry kinetic energy as the molecules in the air vibrate backwards and forwards. When you heat a saucepan of water over a gas flame, the __________ energy from the flame is transferred to the water molecules. The temperature of the water __________ because the water particles are moving faster and have greater kinetic energy. The kinetic energy of a body depends on both its __________ and its velocity. The term *velocity* refers to the __________ of a body when measured in a straight line motion. If a car and a large truck are both travelling at the same velocity, then the truck has a __________ kinetic energy as it is larger. If two bodies of the same mass (m) are moving at different velocities, then the one with __________ velocity has a greater kinetic energy. This occurs because the kinetic energy of a body depends on the square of its __________ (v). Mathematically this can be expressed by the equation: $KE = \frac{1}{2}mv^2$

2 Elastic potential energy is another example of __________ energy. When you stretch an elastic band, you are using energy to uncoil the __________ molecules in the band. While the band is stretched it is storing energy in the form of elastic potential energy. If you remove the force, the band __________ back and releases this energy (similar to the way bungee jumpers stretch the elasticised cable as they fall). When they reach the lowest point and stop, the cable is storing __________ elastic potential energy. As they rise back up, the elastic potential energy is reconverted back into __________ energy. The metal springs inside your bed mattress become __________ when you lie on the bed but they expand again when you move or get off the bed. The compressed springs __________ elastic potential energy.

3 Chemical potential energy is another very important example of stored energy. Chemicals such as glucose contain considerable potential energy in the chemical __________ of their molecule. When we eat foods that contain glucose, chemical reactions occur in which bonds are __________ and new bonds are formed. The chemical potential energy is being __________ into other forms to power your body as well as maintain your body temperature. In a food

chain the solar energy that is absorbed by __________ is stored in the chemical bonds of the molecules that are synthesised. This stored energy in a plant's tissues may be stored for very long periods of time. For example, coal is the __________ remains of ancient plants. When we burn the coal in a furnace we are releasing the __________ energy that has been stored for millions of years.

4 Energy can be stored in a body due to its __________ relative to the surface of the Earth. This type of stored energy is called gravitational potential energy. A dam stores a large mass of water and this water has a high gravitational potential energy. The following photo shows some water in a dam being released. In this case the gravitational potential energy is converted into the kinetic energy of the moving water.

When you lift a heavy weight off the floor, you have to use energy from your __________ to raise the heavy weight against the __________ of gravity. The energy used is now stored as __________ potential energy in the body due to its height above the floor. If you were to drop the body, its gravitational potential energy would be __________ into kinetic energy as it fell. The gravitational potential energy of a body of mass (m) that is raised to a height (h) above the ground is given by the formula: GPE = mgh

In this equation the g stands for the gravitational __________ of the Earth. Its value near the ground is 9.8 m/s^2. The equation shows us that gravitational potential energy depends on both mass and __________. If we were to measure the gravitational potential energy of a raised body on the Moon's surface then g will also change. The __________ of gravity on the Moon is about one-sixth of that on the Earth's surface.

Answers **1** kinetic; molecules (particles); damage; heat; rises (increases); mass; speed; greater; greater; greater; velocity **2** stored; long; contracts; maximum; kinetic; compressed; store **3** bonds; broken; converted; plants; fossilised; chemical **4** position; muscles; force (pull); gravitational; converted; acceleration; height; strength

KINETIC AND POTENTIAL ENERGY

Energy

REVISION SUMMARIES

1 Our roads have speed signs to limit the **maximum speed** for safe driving. The faster a car travels, the greater the damage it can cause if it crashes into another vehicle, pedestrians or a stationary object such as a tree or pole. It has been shown that a car travelling at 20 m/s will cause more than twice the damage than one that is travelling at 10 m/s if it crashes. This is because the faster car has more **kinetic energy** than the slower car. A truck that is travelling at 10 m/s will cause more damage if it crashes than a car travelling at the same speed.

Moving air has kinetic energy. Sound energy is transmitted through the air in the form of a wave. When a sound is produced (e.g. from a trumpet), the particles of air push against each other. The air particles move backwards and forwards when sound energy is transmitted. This process continues as long as the trumpet continues to be played.

Kinetic energy (KE) can be calculated from a knowledge of the mass (m) of the body and its velocity (v), using the following formula:

$$KE = \frac{1}{2}mv^2$$

Kinetic energy is measured in **joules** (J), mass is measured in **kilograms** (kg) and velocity is measured in **metres per second** (m/s). The velocity of a body (object) is its speed when measured over a straight course.

Example: Compare the kinetic energy of a 1000-kg body when travelling at 10 m/s and 20 m/s.

Answer: At 10 m/s, $KE = \frac{1}{2} \times 1000 \times 10^2 = 50000$ J.

At 20 m/s, $KE = \frac{1}{2} \times 1000 \times 20^2 = 200000$ J.

Therefore, the kinetic energy of the body at the higher speed is four times greater than it was at the lower speed.

2 Another form of potential energy is called **elastic potential energy**. Elastic potential energy is stored when a spring is stretched or compressed. When the spring is released, it can do work and apply a force. Consider the springs that support and stretch the canvas fabric of a trampoline. If a person jumps down on the trampoline, the springs will be **stretched** and temporarily store energy as elastic potential energy. This stored energy is released when the springs un-stretch and return to their normal length. This leads to the person being pushed back into the air. In order to fire an arrow, the string and bow are pulled back and the bow is stretched. On release the bow returns to its normal shape and the energy released is used to propel the arrow forward.

The following diagram shows the elastic potential energy in the compressed spring of a toy gun. As the spring expands it pushes on the ball and accelerates it. Its velocity increases while the ball is being pushed by the expanding spring. Thus the elastic potential energy is converted to kinetic energy.

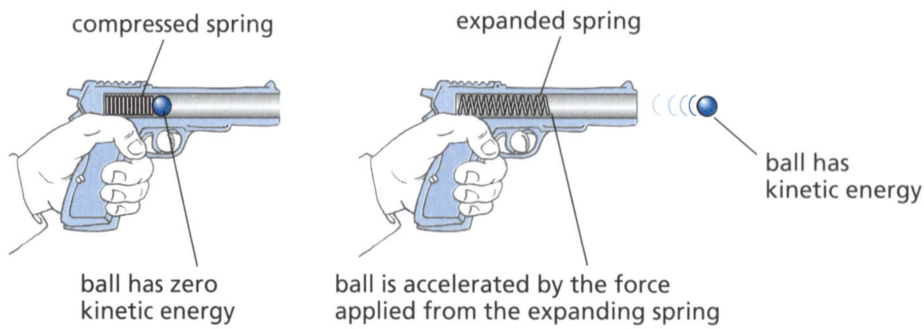

3 Another example of stored energy is **chemical potential energy**. When we eat food, we are consuming energy that is stored in the molecules of food. This energy is then converted to usable forms of energy that will power our muscles and keep our bodies warm. Some foods contain larger amounts of chemical potential energy than others. Fatty meals contain more energy than less fatty foods. Chemical potential energy is also stored in fuels such as petrol and diesel. Inside a car engine, these fuels undergo combustion and some of the stored energy is converted to kinetic energy to power the car. Similarly with rocket engines: the stored energy in the **fuel** (hydrogen) and the **oxidiser** (oxygen) is partly converted to kinetic energy to propel the rocket into the sky.

4 Energy can also be stored in a body. One way to store energy is in the form of **gravitational potential energy** (GPE). If a wooden pole is to be hammered into the ground, a heavy weight (called a pile driver) is raised above the pole and allowed to drop onto the top of the pole. The greater the height the weight is raised the further it will drive the pole into the ground. If a heavier weight is used then it will drive the pole further into the ground than a lighter weight dropped from the same height.

Gravitational potential energy can be calculated using the following formula:

$$\text{GPE} = mgh$$

Like kinetic energy, gravitational potential energy is measured in joules and mass is measured in kilograms. The height (h) of the mass above the ground is measured in metres. The strength of the gravity field (g) is measured in metres per second per second (m/s^2). On the Earth's surface g has a value of 9.8 m/s^2; g is also called the acceleration due to gravity.

Example: Compare the gravitational potential energies of a body (A) of mass 10 kg at a height of 10 m to a second body (B) of mass 20 kg at a height of 5 m. Assume g is 9.8 m/s^2.

Answer: Body A: GPE = $10 \times 9.8 \times 10 = 980$ J
Body B: GPE = $20 \times 9.8 \times 5 = 980$ J
Thus, both bodies have the same gravitational potential energies.

Checklist

Can you:

1 *Identify the two variables that change the kinetic energy of a body?* ☐
2 *Explain how springs and elastic bands store elastic potential energy?* ☐
3 *Explain how matter can store chemical potential energy?* ☐
4 *Identify the three variables that change the gravitational potential energy of a body?* ☐

KINETIC AND POTENTIAL ENERGY

Energy

REVISION TEST

1 The following diagram shows two experiments that were conducted in a school laboratory.

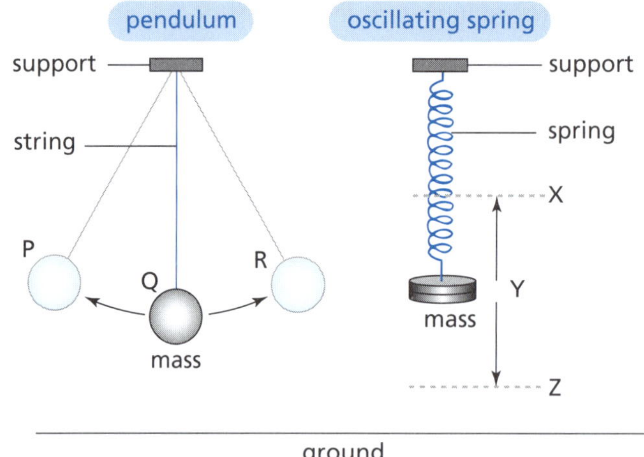

a In the first experiment, a pendulum bob is pulled to the side and the mass allowed to oscillate. Use the letters in the diagram to identify the locations where:
  i kinetic energy of the bob (mass) is at a maximum (1 mark)
  ii gravitational potential energy is at a maximum. (1 mark)

b In the second experiment a mass attached to a spring is pulled down to stretch the light spring. The mass is then released and it oscillates vertically. Use the letters in the diagram to identify the locations where:
  i kinetic energy of the mass is at a maximum (1 mark)
  ii elastic potential energy is at a maximum *Hint 1* (1 mark)
  iii gravitational potential energy of the mass is at a minimum. (1 mark)

2 Examine the following table which lists the energy content of some common foods and information about the carbohydrate, fat and protein content. Assume the remainder is water.

Food (100-g serve)	Energy content (kJ)	Carbohydrates (g)	Fat (g)	Protein (g)
milk (whole)	290	5	4	3
margarine	3000	3	80	0.5
apple	250	15	0.5	0.3
rice (white)	1520	80	0.3	8
minced beef	1600	0	30	22
egg (boiled)	670	0.7	12	13
potato (boiled)	350	18	0.1	2
cheese (cheddar)	1700	2	32	25
corn flakes	1600	85	1	7
bread (white)	1160	50	3	8

a Identify the type of energy stored in food. (1 mark)

b i Identify the food that has the greatest energy content per 100-g serve. (1 mark)
  ii Explain why this food has such a high energy content. (1 mark)

c Explain why an apple has the lowest energy content per 100-g serve. *Hint 2* (1 mark)

d Calculate the total energy content of the following meal. (6 marks)
1 boiled egg   2 × 50 g slices of white bread   25 g margarine
2 × 100 g apples   150 g whole milk

3 Complete the table by recording the type of energy that corresponds to the category in column 2. Choose from the following list: kinetic energy; gravitational potential energy; elastic potential energy; chemical potential energy. (8 marks)

Example	Category	
a	stretched rubber band	
b	E10 petrol	
c	water in a dam	
d	snow avalanche	
e	ocean wave	
f	wound-up spring	
g	parachutist about to jump	
h	compressed air in truck brakes	

4 Calculate the kinetic energy of a 100-g mass that is moving with a velocity of 20 m/s. (2 marks)

5 A body of mass 1 kg has a kinetic energy of 450 J. Calculate its velocity. *Hint 3* (2 marks)

6 A 200-kg rock is resting at the edge of a cliff on the Moon. The rock is 150 m above ground level. Calculate its gravitational potential energy given that the value of the acceleration due to gravity is 1.6 m/s^2. (2 marks)

7 A roller-coaster at a fairground is shown in the following diagram.

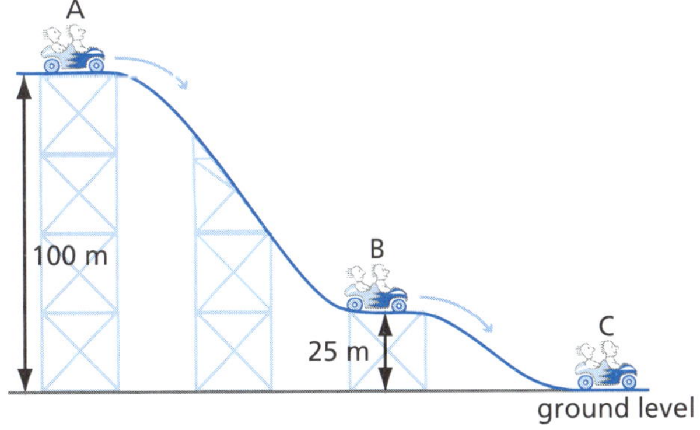

a Compare the relative values of the gravitational potential energy of the roller-coaster at A, B and C. (3 marks)

b At what location will the roller-coaster be travelling at the greatest speed? (1 mark)

Hint 1: There are two locations where the elastic potential energy is at its maximum.
Hint 2: What is the major component of an apple?
Hint 3: Rearrange the equation and make velocity the subject.

Your Feedback

$\frac{\square}{33} \times 100\% = \square\%$

PAGES 198–199
PAGE 212

ENERGY TRANSFORMATION

Energy

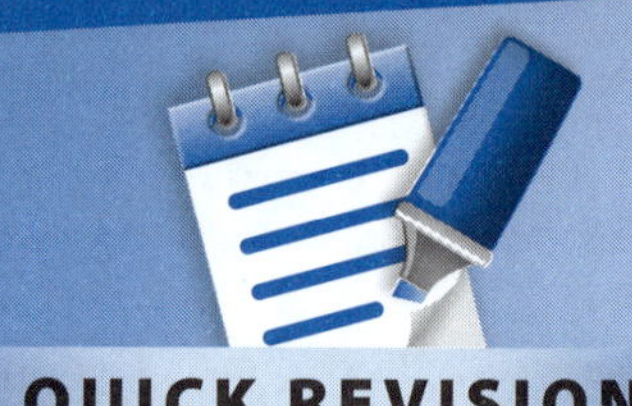

QUICK REVISION

1 The Sun is a yellow-white star that generates its energy via __________ fusion by converting hydrogen to helium. The Sun __________ vast amounts of energy into space. Some of this energy is __________ by the Earth. The atmosphere, lithosphere and hydrosphere absorb this energy and the Earth is warmed. The energy to drive the water __________ comes from the Sun. Water and water vapour absorb large amounts of infra-red __________ and convert it into __________. Green plants use visible radiation to carry out photosynthesis. In this process the solar energy is converted into chemical __________ energy. The transformation of energy from one form to another occurs in every part of the natural and built environment. For example, electrical energy is __________ into kinetic energy in suburban trains. When drums are played, __________ energy is converted into sound energy. When a battery is used to power a torch, chemical potential energy is transformed into electrical energy and then into __________ energy. Energy can also be transferred from one place to another. When a hot piece of metal is placed into a container of cold water, heat energy is transferred __________ the metal __________ the water.

2 Heat energy is released to the surroundings or __________ from the surroundings in various chemical reactions. When magnesium metal reacts with sulfuric acid, the metal dissolves and hydrogen gas is released. The acid becomes __________ in this reaction. This heat energy has been produced by changes in chemical potential energy as __________ are converted to products. When heat is released, the products have __________ chemical potential energy than the reactants. Thermometers can be used to measure the __________ in temperature.

3 The burning of fossil fuels generates large amounts of __________ energy. Petrol, diesel, coal, __________ gas and hydrogen are useful fuels. In Australia, most of our electricity is generated by the combustion of __________ or natural gas (methane). Vast coal fields are found in New South Wales, Queensland and __________ and large reservoirs of natural gas are located off the coast of Western Australia. When these fuels undergo combustion, the heat released is __________ to water which is converted to steam. __________ steam is used to turn the blades of a __________ which generates electricity. In this process the __________ energy of the rotating turbine is transformed into electrical energy.

4 Motors are used in many different appliances in our homes. These include electric drills, fans and __________ cleaners. These machines transform electrical energy into __________ energy. A motor consists of electrical wires that are wound on a steel rotor. The rotor lies between the __________ of an electromagnet. When an electric current flows through the wires, the rotor starts to __________. You can investigate how fast the motor spins as a function of the __________ applied. The greater the voltage or current, the __________ the motor spins. The motor is warm after use. This indicates that __________ is also produced.

5 The law of __________ conservation states that energy cannot be __________ or destroyed. When energy is __________ or transformed, some of this energy is wasted. Often this wasted energy is in the form of __________ or sound. Unused energy is not destroyed. It is still there but in a different but non-useful form. When machines are designed, it is important to __________ this wasted energy.

Answers **1** nuclear; radiates; absorbed; cycle; radiation; heat; potential; transformed (converted); kinetic; light; from; to **2** absorbed; warm (hot); reactants; less; change **3** heat; natural; coal; Victoria; transferred; pressurised; turbine; kinetic **4** vacuum; kinetic; poles; spin (rotate); voltage; faster; heat **5** energy; created; transferred; heat; reduce

1 The Sun generates vast amounts of energy by **nuclear fusion** reactions involving hydrogen. Large amounts of this energy travel through space in the form of **radiant energy** (often called electromagnetic radiation). The Earth absorbs a very small amount of this energy, although some of the incident radiation is reflected back into space. Green plants absorb some radiant energy and through a process called **photosynthesis** convert this energy into chemical potential energy. Molecules such as **glucose** store this chemical potential energy in their chemical bonds. Bodies of water such as the oceans absorb large amounts of radiant energy and convert it into heat energy. Although the Earth has its own internal heat, it is the Sun's energy that powers the water cycle and the carbon–oxygen cycle. The energy for winds also comes from the Sun. All these processes involve various types of **energy transfer** and **transformations**. In the previous section we have examined some energy transformations. For example, gravitational potential energy is transformed into kinetic energy when a body falls under gravity, while elastic potential energy is transformed into kinetic energy when an arrow is fired from a bow.

2 The transformation of chemical potential energy into heat energy is commonly observed in **chemical reactions**. Consider the experiment shown in the diagram below. Drops of water are added to a test tube containing some plaster of Paris. This white powder is composed of a chemical called calcium sulfate. The temperature of the powder is monitored and large amounts of heat are produced during the reaction. In this reaction the calcium sulfate becomes **hydrated** and chemical potential energy is transformed into heat energy.

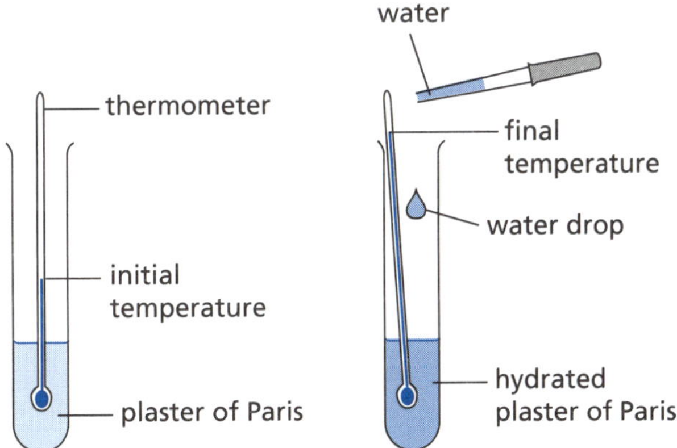

3 When coal is burnt in a **furnace**, chemical potential energy is converted to heat energy. Some of this heat energy can be used to generate electricity at a **power station**. The heat is first used to boil water and turn it into high-pressure steam. This high-pressure steam is storing elastic potential energy which can be transformed into kinetic energy as it is released. The kinetic energy of the steam particles is transferred to the rotor blades of a turbine. As the turbine rotates, kinetic energy is transformed into electrical energy.

4 **Electrical energy** is used to power many machines in our homes and workplaces. Various energy transformation processes occur in different machines. The following diagram shows a simple school experiment in which electrical energy is transformed into kinetic energy in a motor. A fan is attached to the axle of the motor and when the tapping switch is depressed, the electric current flows and the fan blades rotate. The speed of the rotation is adjusted by altering the voltage of the power supply or changing the current via the variable resistor slide. After the

(cont.)

apparatus has been working for some time, the motor and resistor become hot. This shows that energy transfer and transformation is not perfect. Some energy is converted to other forms and wasted.

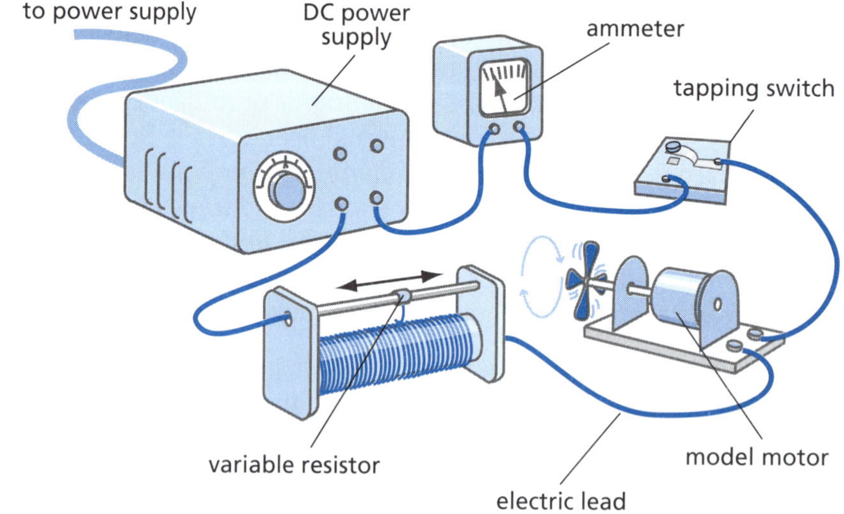

5 An important principle in science is the **law of energy conservation**. This law states:
In all processes, energy is neither created nor destroyed but simply converted into new forms.

Consider the example of coal combustion to generate electricity as discussed in point 3 above. When coal is burnt in a furnace, 100 J of chemical energy in the coal is transformed to finally produce about 30 J of electrical energy. The remaining 70 J is wasted mainly as heat energy in the various stages of the electrical generation process. Friction is a common way in which energy is lost as heat. So lost, or wasted, energy is **not destroyed**. It is simply in a form we can't readily use. Eventually most of this heat energy will be radiated back out into space.

Checklist

Can you:

1. *Explain how the Sun's energy is converted into various forms on Earth?* ☐
2. *Describe an example of the transformation of chemical potential energy into heat energy?* ☐
3. *Describe the energy transformations that occur when coal is combusted to generate electricity?* ☐
4. *Describe an experiment in which electrical energy is converted into kinetic energy?* ☐
5. *State the law of energy conservation?* ☐

ENERGY TRANSFORMATION

Energy

REVISION TEST

1 The diagram below shows two experiments being conducted in the school laboratory.

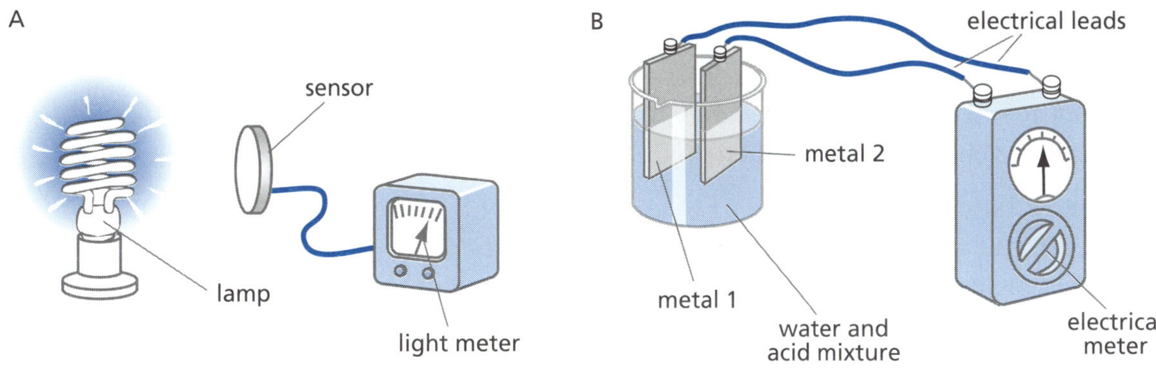

- **a** In each case, state the energy transformation processes that are being investigated. (2 marks)
- **b** In experiment A the student placed the lamp at 20 cm from the sensor and then at 100 cm from the sensor. Predict and explain how the reading on the light meter may change. (2 marks)
- **c** Identify two variables that will alter the current measured by the electrical meter in experiment B. *Hint 1* (2 marks)

2 Energy is transferred from source to receiver. The ocean and the atmosphere are energy receivers because they warm up when they receive energy from the Sun. For each of the following sources of energy, name one energy receiver.

- **a** electric guitar (1 mark)
- **b** gas oven (1 mark)
- **c** cyclone (1 mark)

3 Frederick is a Year 8 student who gets into trouble for playing his CDs too loudly.

- **a** Predict whether loud music has more energy, less energy or the same energy as soft music. Justify your prediction. (2 marks)
- **b** What is the energy source for the CD player? (1 mark)
- **c** Is more, less or the same amount of energy used to play the CD loudly as is used to play it softly? (1 mark)

4 A student lights a Bunsen burner. Using oven gloves, she holds a hollow metal pipe over the Bunsen flame and gradually lowers the pipe. A low noise is initially heard before the flame is inside the pipe. The noise increases to a loud roaring when the flame is actually inside the pipe. The pipe walls become very hot.

- **a** Identify the energy transformation that occurs in this experiment. (1 mark)
- **b** Suggest one possible explanation for the production of the roaring sound. (1 mark)
- **c** Propose a further experiment that could be carried out to test your explanation in part b. *Hint 2* (1 mark)

(cont.)

5 Give an example of a process in which the following energy transformations occur.

- **a** chemical potential energy to heat energy (1 mark)
- **b** kinetic energy to electrical energy (1 mark)
- **c** solar energy to chemical potential energy (1 mark)
- **d** electrical energy to sound energy (1 mark)
- **e** kinetic energy to electrical energy (1 mark)

6 True or false?

- **a** All the energy from the Sun is absorbed by the Earth. (1 mark)
- **b** Fat molecules store chemical potential energy in their bonds. (1 mark)
- **c** When a ball is thrown into the air, kinetic energy is transformed into gravitational potential energy. (1 mark)
- **d** When calcium oxide is mixed with water, the mixture gets hot. In this example elastic potential energy is converted to chemical potential energy. (1 mark)
- **e** Some energy is destroyed when coal undergoes combustion in a power station. (1 mark)

7 Machines are energy converters or transformers. They can transform one type of energy into another. Name an example of an energy converter in each of the following energy transformations.

- **a** light energy to electrical energy (1 mark)
- **b** electrical energy to kinetic energy (1 mark)
- **c** electrical energy to light and sound energy (1 mark)
- **d** chemical potential energy to kinetic and heat energy (1 mark)
- **e** electrical energy to gravitational potential energy (1 mark)

8 On a future space city on the Moon, solar cells absorb 100 units of energy into their storage batteries. This is used to produce 43 units of electricity for the city. By the time the electricity reaches the power points, there are 35 units of electricity available to charge the batteries of the electric cars. The cars only get 20 units of energy from their fully charged batteries to produce movement.

Draw a flow chart of these energy transfers and transformations. Include the energy wasted as heat in your flow chart. *Hint 3* (7 marks)

Hint 1: Variables are factors that influence the result of the experiment
Hint 2: How can you change the speed of the hot moving air?
Hint 3: Ensure that each box (after the solar cells) of the flowchart has an energy value inserted.

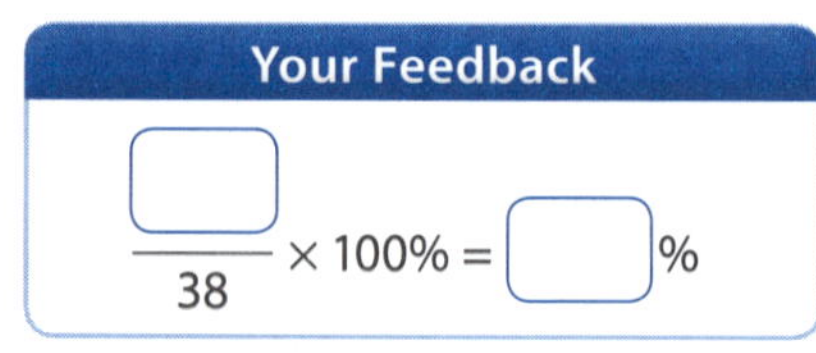

HEAT ENERGY

Energy

QUICK REVISION

1 Some people think that ________ and temperature are the same thing. But this is not so. While heat and temperature are related, they are different ideas. This notion is often confused because in everyday experience when you heat something up, its temperature ________. And when you ________ something down, its temperature falls.

Temperature is not ________, it is a measure of it. It is a number that is related to the average kinetic energy of the ________ of a substance. The molecules of a substance are in constant, random and rapid ________. Some particles may be moving or vibrating faster than other particles at any particular time. If they get knocked or bumped then the ________ of these particles can change. These changes occur very quickly and by chance, so scientists talk about average speeds. If their average motion increases, the ________ rises; if the average motion ________, the temperature drops. Temperature is measured with a device called a ________. We use the ________ scale where 0 °C represents the freezing (________) point of water, while 100 °C represents the ________ point of water.

Heat is energy, and is measured in joules. If heat is added to a substance, that energy can be used to increase the ________ energy of the molecules. If the average motion of the particles increases, then there is an increase in the ________. At certain temperatures any added heat could be used to break the bonds between molecules resulting in a change of state. As their average ________ does not change, there is no change in temperature. A good example is melting ice. At 0 °C ice begins to melt, forming ________ water. Even though ice is melting to water, the temperature remains at ________ until melting is complete.

2 There are three main ways that heat is ________ in our environment: conduction, convection and radiation. Conduction is the transfer of heat between objects that are in direct ________ with each other. Metals are examples of good thermal (________) conductors. These include copper, iron, steel and aluminium. Poor conductors, often called ________, include wood, cloth, paper, air and styrofoam.

3 Convection is the main way that heat moves through liquids and gases, as these can ________. When a gas or liquid is heated it ________ as it warms, becoming less dense. This causes it to rise. Eventually some of the heat can be transferred elsewhere and the gas or liquid cools. On cooling it becomes ________ and falls. This rising and falling creates a ________ current and transfers heat. Convection accounts for a pool being warmer at the surface than at the ________, for wind currents and why higher floors in a building are warmer than lower floors.

4 Sunlight is a form of ________ that is radiated through space to our planet without the aid of fluids or solids. The energy travels as electromagnetic ________ through space. When these waves come in contact with an object, they are changed into ________. Our planet's atmosphere is ________ by this radiation. Other examples of where radiation can heat up objects include an open fire and a microwave oven.

Answers 1 heat; rises; cool; energy; molecules (particles); motion; speed (velocity, motion, movement); temperature; decreases; thermometer; Celsius; melting; boiling; kinetic; temperature; motion; liquid; 0 °C 2 transferred (moved around); contact; heat; insulators 3 flow (move); expands; denser; convection; bottom 4 energy; waves; heat; warmed

1 **Temperature** is a measure of the average kinetic energy of the particles that make up a substance. Higher temperatures mean that the particles are moving, vibrating and rotating with more energy. Temperature is measured using a thermometer in units called degrees Celsius (°C).

The diagram shows two gases at different temperatures. On average, the particles in the left-hand container are moving more slowly than those on the right, so the gas is at a lower temperature. (The longer the arrow is, the higher the speed of the particle.)

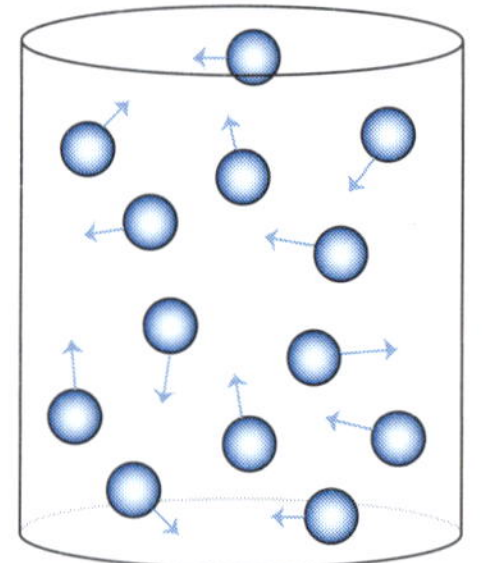

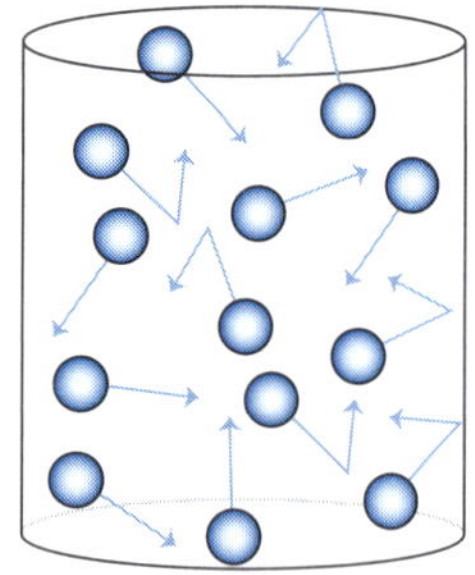

Heat is a form of energy and is measured in joules (J). It is the total energy of molecular motion in a substance. Energy is added to a substance when heat is added. This added heat energy often increases in the **kinetic energies** of the molecules, and so the temperature rises. If heat energy changes the state of the substance, say by melting or boiling it, then the added energy is used to break the bonds between the molecules rather than increasing their kinetic energy. So while the substance is melting, or boiling, the temperature doesn't increase even though heat is still being added. This heat is called **latent** (hidden) heat.

Heat energy depends on how fast the particles are moving, the number of particles (that is, the size or mass of the substance) and what kind of particles are in the object. Temperature does not depend on the size or type of object. For instance, the temperature in a cup of water might be the same as the temperature of the water in a bathtub, but the tub has more heat because as there is more water and so more total thermal energy.

If two objects at the same temperature are in contact, there is no overall transfer of energy between them because the average energies of the particles in each object are the same. But if the temperature of one object is higher than that of the other object, heat energy will transfer from the hotter to the colder object until both objects reach the same temperature.

2 **Conduction** is the transfer of energy through matter from particle to particle. For example, a metal spoon in a cup of hot coffee becomes warmer because the heat from the coffee is conducted along the spoon. The flow of heat by conduction occurs when atoms and molecules in the substance collide and transfer kinetic energy. Faster-moving particles collide with the slower ones. In such collisions, the faster particles lose some speed and the slower ones gain speed. And so the faster particles transfer some of their kinetic energy to the slower ones. Conduction is most effective in solids. The following diagram shows conduction of heat along a metal rod.

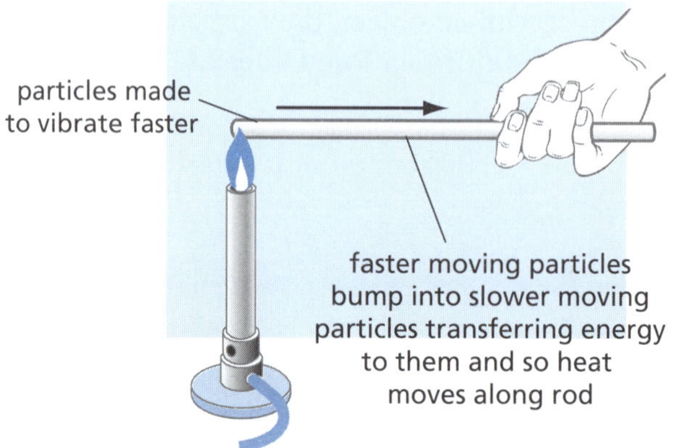

Different materials transfer heat by conduction at different rates. A metal ladle stirring hot soup is a good heat conductor, as are many metals, while the plastic or wooden handle is a poor heat conductor (good insulator). Home **insulation** is a poor thermal conductor.

3 **Convection** is the flow of heat through the movement of matter from a hot region to a cool region. Convection transfers heat energy in a gas or liquid by the movement of currents. The heat moves with the fluid. This type of current is demonstrated in the figure below.

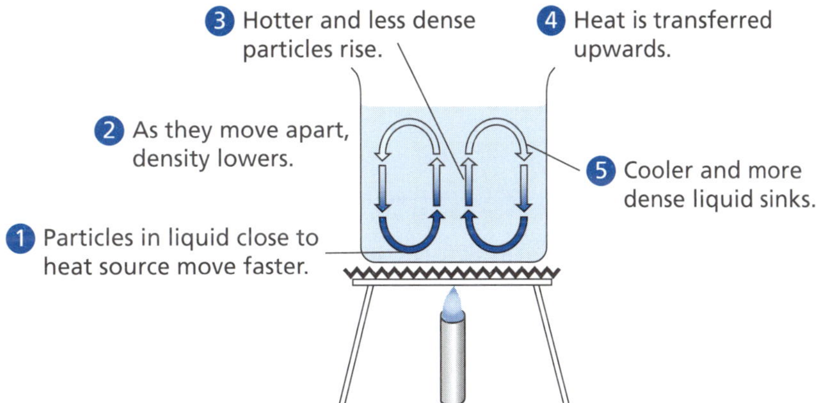

4 **Thermal radiation** is the way energy, as electromagnetic radiation, is emitted by a heated surface. Radiation does not require a medium to carry it, and travels out in all directions to its point of absorption. The intensity and distribution of radiant energy depends on the temperature of the emitting surface, and on the nature of the surface as well. Objects that are good emitters are also good absorbers. A black surface is an excellent emitter as well as an excellent absorber. If the same surface is silvered, it becomes a poor emitter and also a poor absorber. The heating of the Earth by the Sun is an example of transfer of energy by radiation, and so is heating a room by an open-hearth fireplace.

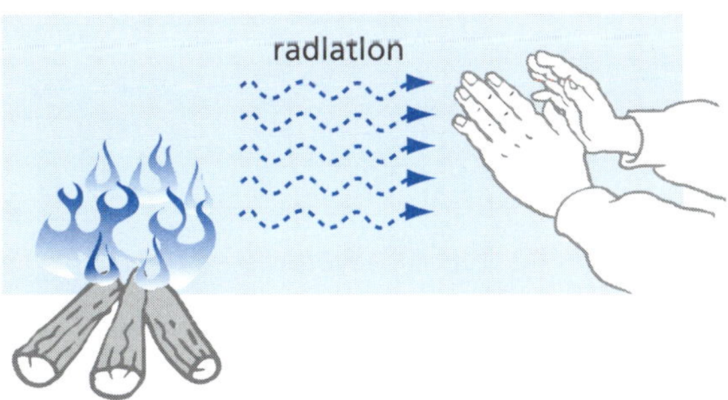

Checklist

Can you:

- **1** *Distinguish between heat and temperature?* ☐
- **2** *Describe and give examples of conduction?* ☐
- **3** *Describe and give examples of convection?* ☐
- **4** *Describe and give examples of radiation?* ☐

HEAT ENERGY

Energy

REVISION TEST

1 **a** What is the difference between temperature and heat? (2 marks)

b In what units is each of these measured? (2 marks)

c The heat content of matter and the temperature of matter depend on a variety of factors. Four of these factors are listed in column 1 in the following table. Copy and complete this table about heat content (column 2) and temperature (column 3) by writing yes if the property depends on the selected factor or no if it does not. (8 marks)

This property depends on the ...	Heat content	Temperature
speed of the particles.		
number of particles.		
size or mass of the particles.		
type of particles.		

2 In the diagram on the right, replace the letters P, Q and R with the correct word from the following list:
conduction; convection; radiation. (3 marks)

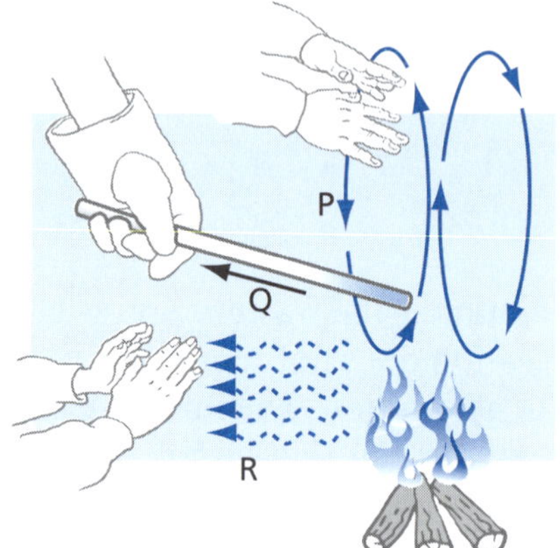

3 Consider the following two diagrams.

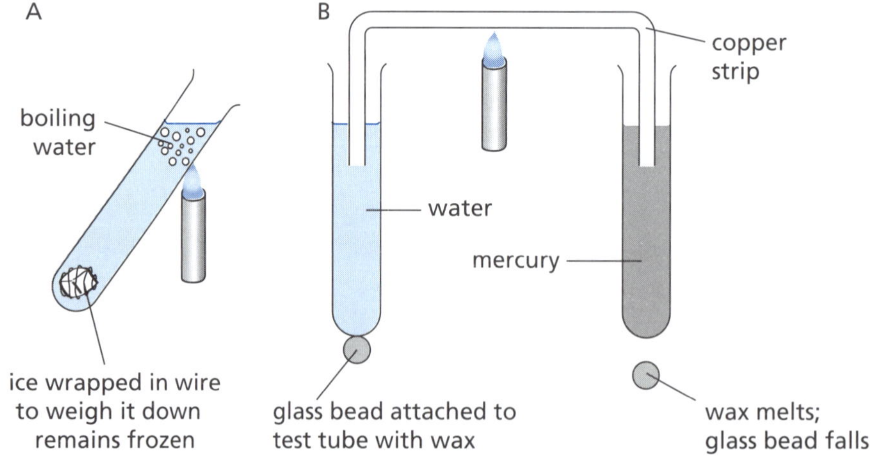

a Explain how A shows that water is a poor heat conductor. *Hint 1* (2 marks)

b Explain how B shows that mercury is a better heat conductor than water. (2 marks)

4 Comment on Jenna's statement:
'Food in the fridge is cold because the refrigerator pumps coldness onto the food to lower its temperature.' *Hint 2* (2 marks)

5 The following nomogram shows how to convert between Celsius temperature (°C) and Fahrenheit temperature (°F).

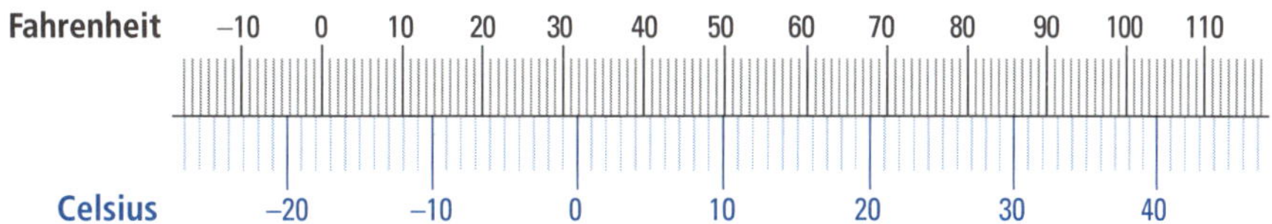

a Water freezes at 0 °C. What is this temperature on the Fahrenheit scale? (1 mark)
b Human body temperature is about 98.6 °F. What is this on the Celsius scale? (1 mark)

6 A screwdriver lies on a table and is at room temperature. If you pick it up by the metal end, it feels colder than if you pick it up by its plastic end. Explain. *Hint 3* (2 marks)

7 Explain the following.
a Feathers and fur keep animals warm, and clothes keep us warm. (2 marks)
b A styrofoam cup can contain boiling tea and yet only a few millimetres away our hand can barely feel this heat. (2 marks)

8 The diagram shows an example of convection currents resulting in sea breezes. Arrange the following in the correct sequence to describe how sea breezes form. *Hint 4* (3 marks)

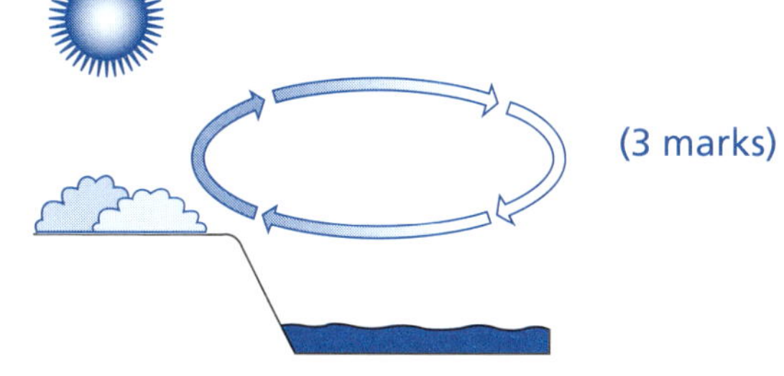

A. Also there is a high pressure area over the water where cooler air falls.
B. This creates a low pressure area over the water relative to the high pressure area over the land.
C. During daytime, water does not heat up as much, or as fast, as land.
D. During the night, water cools off more slowly than land does so the air above the water is slightly warmer than over the land.
E. So during the day, the air above the water will be cooler than that over the land.
F. Now breezes blow from the land to the water.
G. This creates a breeze blowing from the water to the land.
H. There is a low pressure area over the land as warmer air rises and expands.
I. This is because water has a larger heat capacity than land, and so can hold heat better.

9 Hot water and steam, both at 100 °C, can damage the skin. Yet the extent of skin burn (scalding) from steam is greater than an equivalent amount of hot water. Explain. *Hint 5* (2 marks)

Hint 1: Study the diagram and determine what it shows.
Hint 2: You need to understand how heat and coldness are related.
Hint 3: Objects lying around a room for some time obtain the same temperature as the air and other objects in the room. The answer to this apparent contradiction has to do with heat conduction.
Hint 4: Arrange these statements in order, beginning with what happens during daytime.
Hint 5: Steam is not hotter than water, if both are at the same temperature. Consider the change from gaseous steam to liquid water.

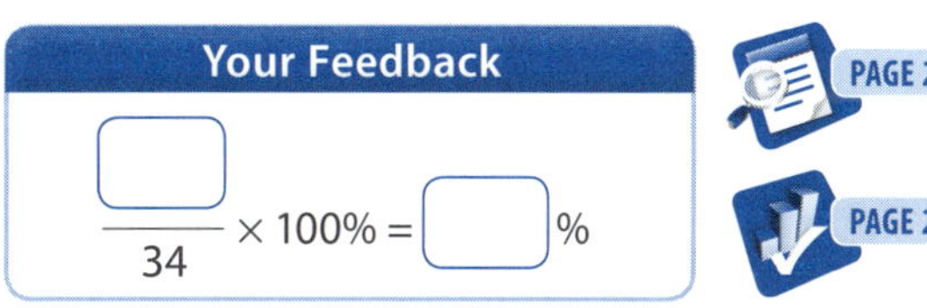

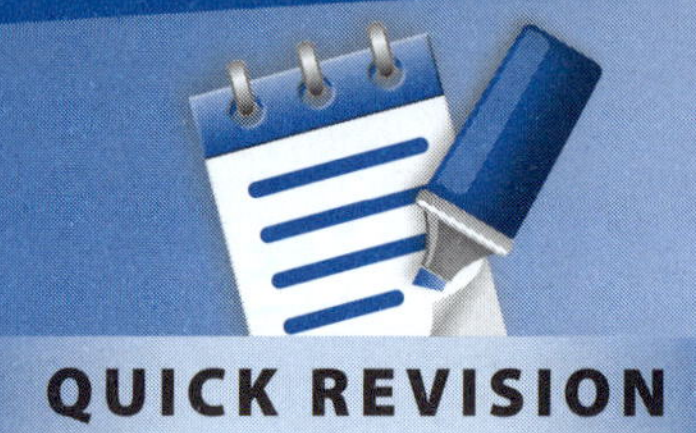

ENERGY EFFICIENCY

Energy

QUICK REVISION

1 Most houses use __________ for their energy needs. Some houses may use piped __________ for cooking and heating, while other houses not near a mains gas supply can have bottled gas delivered. In some homes, __________-fired heaters are used to warm the house during winter.

Heating __________ uses around one-third of the energy used in most homes. Space heating and cooling require a lot of power and use around a quarter of the home's __________ supply.

Using appropriate building materials to __________ the home from extremes of outside temperatures can even out the temperature, thereby reducing __________ and cooling costs. Electrical appliances and equipment also use __________. The amount of time each appliance is used, and its efficiency, have a big impact on how energy __________ the home is. For example, turning off power at the switch when the appliance (television, computer) is not in use saves __________ that would be drawn when it is in stand-by mode. Similarly, turning lights __________ when leaving a room can also save on power costs. Replacing older incandescent __________ with compact fluorescent light bulbs produces three times the light for the same amount of power.

2 Cars are the main method of __________ in Australia, and many households have at least one vehicle. One of the biggest expenses in running a car is the __________ they use. As fuel costs keep increasing, people are starting to consider more fuel-__________ cars. One way to do this is to buy __________ cars as their running costs are less. Currently petrol and diesel-fuelled cars emit __________ gases, and this contributes to climate change. For every litre of petrol used in a motor vehicle, around 2.3 kilograms of .... dioxide (CO_2) is released into the atmosphere. Diesel-powered cars emit 2.6 kg CO_2/L of fuel used.

Many car manufacturers are examining ways to lessen the running costs. One of these is to make cars more efficient. Efficiency refers to the process of converting the chemical __________ energy contained in fuel into useful kinetic energy or __________; that is, allowing the car to cover a greater distance for the same amount of fuel used. Fuel efficiency depends on many features of a vehicle, including its engine design, aerodynamic drag and __________. There have been __________ in all areas of vehicle design in recent decades.

3 Solar energy or solar power uses the energy of __________ to generate electricity, or heat water, for homes, businesses and industry. Sunlight is a renewable and sustainable __________ source. In other words, it doesn't run out and supply can be __________. There is a move towards using clean, renewable energy such as solar, wind, geothermal steam or __________-electricity.

Solar energy can be used in homes in two ways.

- Photovoltaic (PV) solar cells or panels directly convert sunlight into __________.
- Solar water heating systems use the Sun to heat an absorber plate in the collector, and this heats the __________ running through tubes inside the collector. The heated water is pumped or flows under gravity into a __________ tank.

Using solar energy in the home can save on the cost of heating water.

Answers **1** electricity; gas; wood (coal); water; energy; insulate; heating; power (energy, electricity); efficient; power (energy); off; globes (lights) **2** transport; fuel (petrol); efficient; smaller (more economical); greenhouse; carbon; potential; work; weight; advances **3** sunlight; energy; maintained; hydro; electricity; water; storage

ENERGY EFFICIENCY

REVISION SUMMARIES

1 Reducing **energy consumption** around the home is important, especially as energy costs are increasing. With the dramatic increase in home appliances in recent years (such as computers, televisions, wi-fi, microwave ovens, cable services, coffee machines, bar fridges) the average house is becoming more energy-hungry. The following are some steps to reduce energy consumption.

- Turn down the thermostat on air conditioners. At the same time, don't heat rooms you are not using.
- The hot water thermostat does not need to be higher than 60 °C, as this is wasteful of power.
- Prevent cold draughts of air since, in an average house, up to 50% of heat can be lost this way. Closing curtains or shutters after dark traps in warm air.
- Wash economically with a full load in the washing machine and at a lower temperature setting. A front-loader washer can save considerable amounts of water and energy compared to top-loading washers.
- Hang clothes to dry on a line instead of using a tumble dryer.
- Taking a shower instead of a bath uses about half the energy.
- Turn off any appliances not being used, and don't use the stand-by function on televisions, stereos, computers or DVD players.
- You can **prevent heat loss** by insulating. Up to one-third of your home heating escapes through the roof.
- Switch off lights when you leave the room.
- Use energy-efficient appliances. In Australia, a star rating system on many electrical appliances tells you how efficient they are. Look for these when buying a new dishwasher, refrigerator, television, microwave or washing machine.

The most energy-efficient light is natural lighting. A well-designed house with north-facing windows, skylights and light tubes lets in light without compromising the warmth inside. There are various types of lights used in homes. As the graph shows, compact fluorescent lamps (CFLs) are more **efficient** than traditional incandescent globes and last longer. As incandescent lamps are inefficient, they are no longer sold in Australia although some specialty use ones will continue to be available.

Comparing efficiency of globes

Type of light: incandescent globe; halogen; low-voltage halogen; magnetic ballast CFL; electronic ballast CFL; electronic ballast tube

Efficiency (light output per unit)

2 Cars consume natural resources (especially **fossil fuels**) more quickly than the Earth can replace them. With increasing petrol costs, motorists are demanding more energy-efficient cars. Many people are downsizing to smaller, lightweight cars as these use less fuel per kilometre travelled. Fuel consumption is the amount of fuel used per unit distance. In Australia this is usually measured in litres per 100 kilometres (L/100 km). The lower the value, the more economic a vehicle is (that is, the less fuel it requires to cover a certain distance).

Vehicle manufacturers and petrol producers are also experimenting with alternative fuels. **Ethanol** blended with petrol is common. E10 fuel refers to 10% ethanol mixed with 90% petrol. **Bioethanol** (ethanol produced from plants such as sugar cane, corn or other agricultural

(cont.)

feedstock) is a form of renewable energy that can be made from very common crops. Another is hybrid vehicles which can use two or more power sources for propulsion. In many designs, a small combustion engine (using petrol) is combined with electric motors. The battery packs can be recharged by connecting to common household electricity. Kinetic energy which would otherwise be lost as heat during braking can be turned into electricity to recharge a battery and this electrical power can improve fuel efficiency.

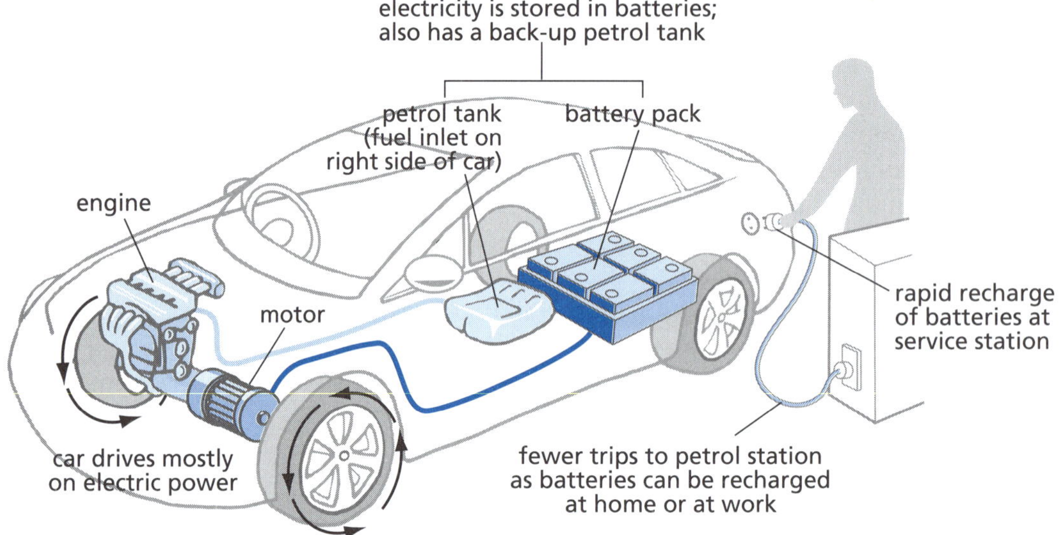

The new generation of **fuel-efficient** cars and hybrids are increasing economy by significant margins. In the next few years totally electric cars will emerge onto the market resulting in cars with zero net carbon emissions.

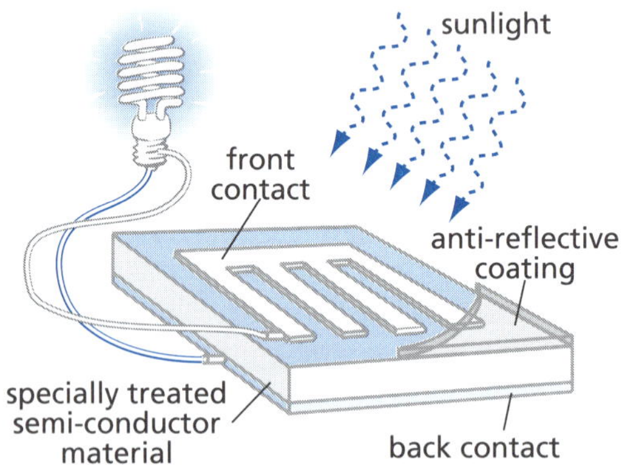

3 **Solar energy** and heat from the Sun have been used by humans since ancient times in different ways. Solar energy technologies in the home include solar heating, where water or air is heated, and solar **photovoltaics** (PV), where solar energy is turned into electricity. Some materials are able to absorb photons of light and release electrons. These free electrons are used to produce an electric current.

Some of the simplest PV cells are used to operate wristwatches and calculators. More complicated systems of panel arrays are used to light houses, for example the number of arrays used determines the total electricity produced. The diagram shows the operation of a basic photovoltaic cell, also called a solar cell.

Checklist

Can you:

1 *State different ways of reducing energy consumption in the home?* ☐

2 *Describe some of the benefits and features of energy-efficient cars?* ☐

3 *Describe how solar energy can be used in the home?* ☐

ENERGY EFFICIENCY

Energy

REVISION TEST

1 The following table shows the percentage energy use in a typical home. A student wishes to draw a sector graph showing this information. Complete the third column in the table (the second column has been done for you) and then draw the sector (pie) graph. (7 marks)

Category	Percentage used	Central angle
water heating	35	$35 \div 100 \times 360° = 126°$
electrical appliances	34	
space heating	24	
cooking	5	
space cooling	2	

2 Comment on the function of each of the following in a house.

a ceiling insulation of dense-pack fibreglass, cellulose or spray foam (1 mark)

b double or even triple-glazed windows (1 mark)

c blinds and curtains across windows (1 mark)

d choosing appliances with a high star energy rating (1 mark)

3 a What is stand-by power? (1 mark)

b Give three examples of devices in the home that use stand-by power. (1 mark)

c It is estimated that stand-by power uses around 3% of the total energy used in a house. Suggest how this can be reduced. (1 mark)

4 The following table shows the annual CO_2 emissions and petrol costs for a year for cars travelling a given distance.

Fuel consumption (L/100 km)	Annual CO_2 emissions (kg)	Petrol use per annum (L)*	Petrol costs per annum**
6		900	$1440
8			
10	3450		
12			

* based on the average vehicle driven 15 000 km in a year ** based on the average petrol cost of $1.60 per L

a What is meant by the unit: L/100 km? (1 mark)

b Complete the column for CO_2 emissions. *Hint 1* (3 marks)

c Why are asterisks (*) used in the heading for the third and fourth columns? (1 mark)

d Describe how 900 L was obtained in the first row. *Hint 2* (1 mark)

e Complete the third column. (3 marks)

f Describe how $1440 was obtained in the first row. *Hint 3* (1 mark)

g Complete the fourth column. (3 marks)

h State one important feature that is shown by this table. (1 mark)

5 a What is meant by a hybrid vehicle? What are the power sources? (2 marks)

b Why are hybrid vehicles becoming more popular? (1 mark)

c Explain what is meant by a plug-in electric vehicle. (1 mark)

d Name two other energy sources being used to drive vehicles. (1 mark)

(cont.)

6 The automotive industry believes using light metals and light alloys for car and truck components is a key way of creating more fuel-efficient cars.

a Why is this so? *Hint 4* (1 mark)

b The CSIRO has developed lightweight car components made from magnesium and aluminium alloys. List three features these alloys must have if they are to compete favourably against current car parts made from steel. (1 mark)

7 One of the most common types of solar-heated hot water system is a roof-mounted flat plate collector and storage tank combination.

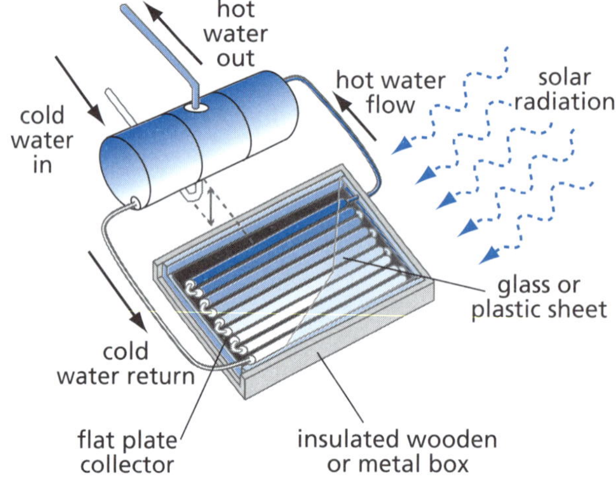

a Describe how this system works. *Hint 5* (2 marks)

b Suggest a reason why cold water enters at the bottom while hot water exits at the top. (1 mark)

8 a Explain what is meant by this statement:
Windows are the weak link, as far as heat goes, in most homes. (1 mark)

b Describe what is being shown in the following diagram. (1 mark)

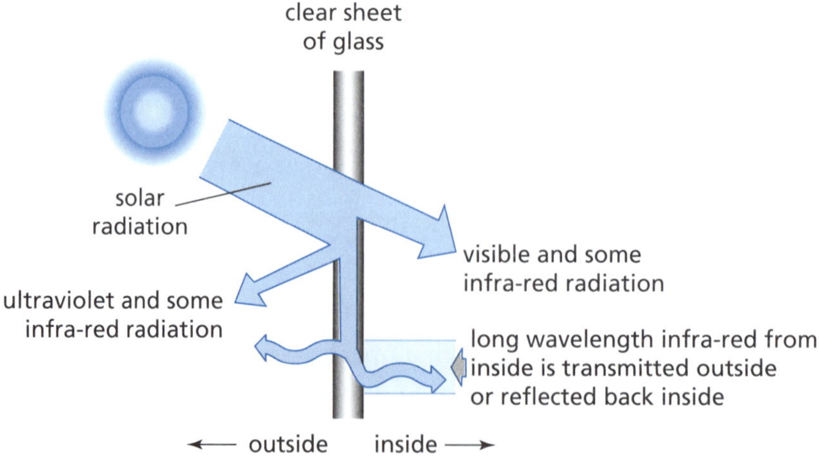

c Why are double-glazed windows better at keeping the heat inside a home? (2 marks)

9 Solar or photovoltaic cells were originally expensive but are now finding many applications on numerous devices and buildings. The following photo shows solar panels being installed on a roof.

a What happens in a solar cell? (1 mark)

b Why would using solar panels to generate electricity be of advantage for homes or farms a long way from the nearest town? (1 mark)

c Electricity from photovoltaic cells has been described as both a clean and a renewable source of energy. What is meant by this? (1 mark)

d List two advantages of using solar panels. (1 mark)

e List two disadvantages of using solar panels. (1 mark)

Hint 1: A car using 10 L/100 km produces 3450 kg CO_2 in a year. Use this to work out the mass of CO_2 for every 1 L/100 km. This will then allow you to calculate the other values.

Hint 2: In order to calculate this value, you need the number in the first column and the value given by the asterisk (*).

Hint 3: In order to calculate this value, you need the number in the third column and the value given by the double asterisk (**).

Hint 4: Consider kinetic energy.

Hint 5: Follow the flow of water around.

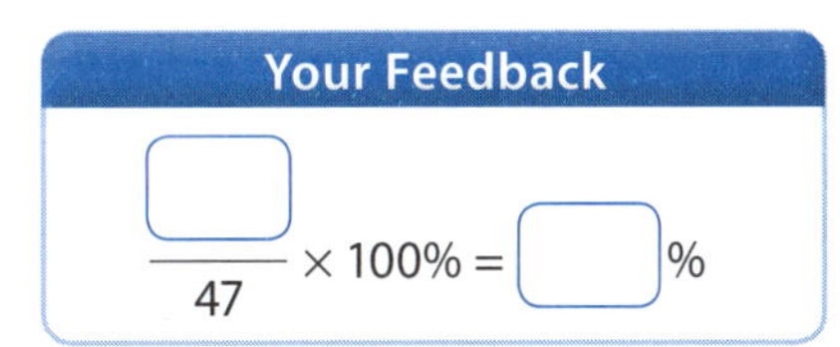

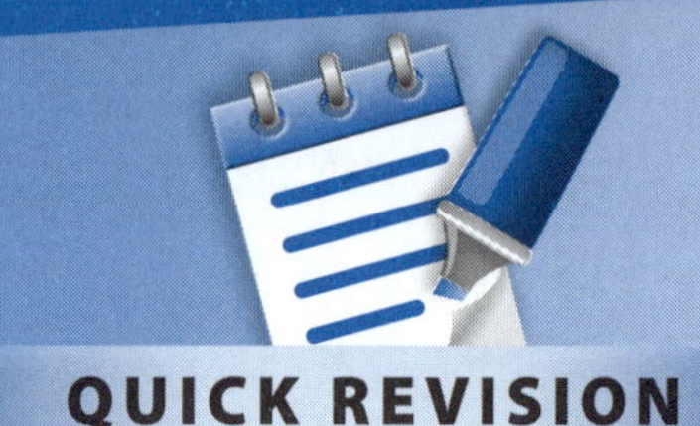

FAIR TESTING

QUICK REVISION

1 The scientific method involves a series of steps to ensure that a first-hand investigation is __________. Such investigations begin when __________ are made that make people curious. A scientist wants to know what caused the event that was observed. Initially the scientist will make various __________ about the event but not all of these inferences may be true. The next step is to make __________ about each inference and then conduct experiments to see if these predictions are borne out. The experiments must be fair and carefully designed to __________ these predictions. If predictions are not supported by the new experiments then new __________ need to be made. Eventually after much testing the scientist may be able to make a __________ about the original event or observation. The generalisation should be __________ for most other related examples. For example, a chemical generalisation is that chloride compounds are soluble in water. Testing has shown that all but a few chloride compounds dissolve in water but there are a few __________ such as silver chloride.

A scientific __________ is a statement that attempts to explain a process or event. Theories are only developed once __________ experiments have been conducted. Once the theory is proposed then __________ it will still continue. If a new experiment fails to confirm an existing theory then a __________ theory may need to be developed or the old one modified. The theory of evolution by natural selection was proposed by Charles Darwin in the nineteenth century. Darwin made many observations in order to develop his theory of evolution. Observations of iguanas (see photo) on the Galapagos Islands were important components of his investigations.

Since then the theory has been modified as __________ knowledge is gained. Experiments in genetics, for example, have helped scientists to explain how evolution can occur. Some theories eventually become laws. Laws are fundamental principles that __________ natural phenomena but even long-standing laws may have to be modified. For example, the law of energy conservation had to be __________ to become the law of mass–energy conservation once Einstein proposed the equivalence of mass and energy.

2 The term *hypothesis* encompasses the ideas of making inferences, __________ and generalisations. Hypotheses are __________ guesses. A hypothesis is a proposal that tries to explain a collection of observations and inferences and it suggests a method of __________ them. Usually hypotheses are not developed until considerable second-hand research is conducted. Libraries and the Internet are helpful here as they may contain __________ to similar observations and experiments by others. It is important that these __________ of

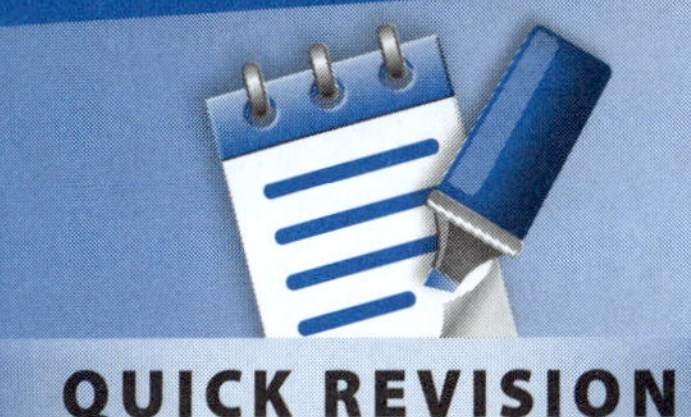

FAIR TESTING

Investigations and problem solving

QUICK REVISION

second-hand data are reliable. Following this research, a scientist is now in a position to create a hypothesis which can then be tested. The tests may not __________ the hypothesis and in this case a new hypothesis will need to be developed.

3 When a scientist conducts a first-hand investigation, they need to list all the __________ that could alter the results. These factors are called variables. Only __________ variable can be investigated at a time. All other variables must be __________ constant. These variables are called __________ variables. Consider the example in which a scientist wants to determine how fast heat is conducted along rods made of alloys of copper and tin. The composition of the alloys is the __________ variable in this investigation. The scientist can choose alloys that are 80% copper, 60% copper and 40% copper. The speed at which heat conducts along the rods has to be measured for each rod and this is the __________ variable. Other variables, such as rod diameter, rod length and heat source need to be held constant and so these are the __________ variables. If the results are plotted on a grid, then the independent variable is plotted on the __________ axis and the dependent variable is plotted on the vertical axis.

In some experiments a control is used. A control experiment is one which is used as a __________ with which other variables are compared. For example, in a test of the effectiveness of a new pesticide to be used on wheat crops, the control will be wheat crops that are __________ sprayed and the test crops will be those that are sprayed with various concentrations of the pesticide. The following photo shows a beetle attacking a crop of wheat.

The concentration of the pesticide will be the __________ variable and the number of dead pests is the dependent variable. Such an experiment would be difficult to control in an open field and so it may be done in a closed __________ such as a greenhouse.

Answers 1 fair; observations; inferences; predictions; test; inferences; generalisation; true; exceptions; theory; many; testing; new; new; explain; modified 2 predictions; not; testing; references; sources; support 3 factors; one; kept (held); controlled; independent; dependent; controlled; horizontal; standard; not; independent; environment

FAIR TESTING

Investigations and problem solving

1 When you investigate a problem scientifically, you take the following steps. These steps are called the scientific method.

- Make and record **observations** using your senses and scientific instruments.
- Seek explanations by making **inferences**. There may be alternative explanations and inferences may not be correct following further investigation.
- Make predictions based on these inferences. Any prediction must be able to be tested.
- Test **predictions** by undertaking new experiments. New experiments may support or not support the predictions. If the prediction is not supported then further investigations need to be planned and conducted.
- Develop a **generalisation** following ongoing investigations. Through continued research and testing, the original inference may be changed until a point is reached when the scientists are confident that a broad statement, called a generalisation, can be made. This generalisation should be true for most related observations.

Most people have heard of scientific theories and laws. These terms have very specific meanings in science. For example, the Big Bang theory is a general statement that attempts to explain how the universe came into existence. It is called a theory as it is consistent with all available evidence. The Big Bang theory can be used to predict new phenomena or explain newly discovered phenomena. Theories in science are backed by a great number of **observations** and **experiments**. The **theory** of the evolution of life by natural selection is also backed by a wealth of evidence and explanations as to why change occurs. Eventually many theories become laws of science. For example, Newton's law of inertia is a law as it explains all known facts about uniform motion. This **law** states that a body will stay at rest or in a state of uniform motion unless acted on by an unbalanced force. This is a fundamental law of physics.

2 The three steps called inference, prediction and generalisation can be combined into one idea called the **hypothesis**. A hypothesis is a broad, general explanation of a collection of observations that predict a result and suggest a way of **testing** it. A hypothesis is different from an inference. A hypothesis is only made after looking at all the available **evidence** and information. Some information comes from background reading or the ideas of other scientists. The results of many other related experiments help to produce the hypothesis. A hypothesis does not try to explain only one observation. It is a more general statement that will explain many separate observations related to the problem. Hypotheses contain only opinions that can be tested. The following diagram illustrates the process of developing and testing a hypothesis.

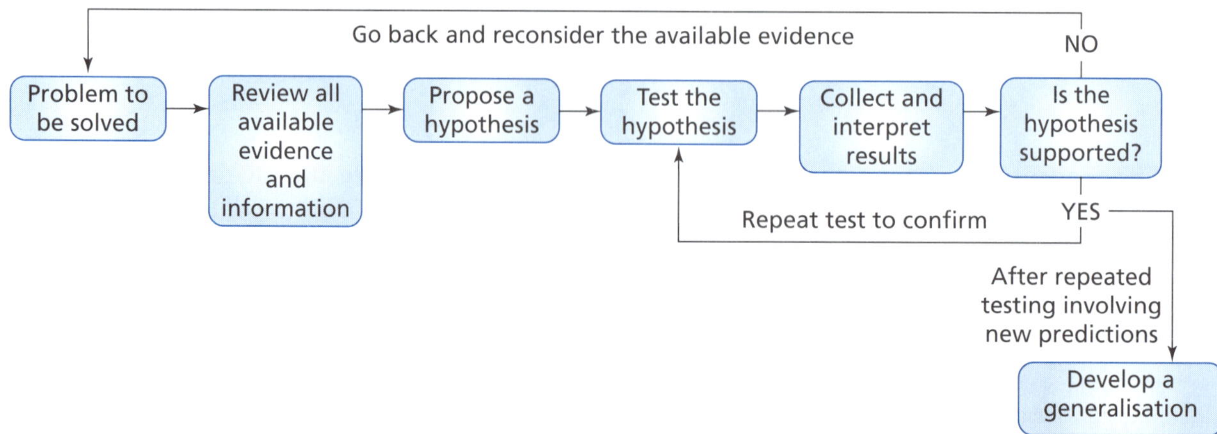

3 There are often many factors that could change the result of an experiment. These factors are called **variables**. In order to investigate a problem experimentally, you must keep all but one variable constant so that only the variable being investigated can affect the result. This principle is known as fair testing. The results from a first-hand investigation are only reliable if the tests are fair.

- **Independent variable.** This is the variable that the experimenter allows to change in a systematic way.
- **Dependent variable.** This is the variable that is measured in the experiment. The dependent variable is affected by the change in the independent variable.
- **Controlled variables.** The controlled variables are the other factors that can alter the result if they are not held constant.

Measurements may only be **reliable** in some experiments if a **control** is used. A control is an experiment that is performed as a comparison with those in which the independent variable is allowed to change.

For example, in an experiment in which a fertiliser is tested to determine whether it makes plants grow faster, the scientist needs to have an equal number of similar plants that are not fertilised. These non-fertilised plants are the control. The rate of growth of the fertilised plants can then be compared with the control plants. The following photo shows seedlings that are being tested with a fertiliser. An equal number of seedlings will act as controls.

Checklist

Can you:

1 *List the stages of the scientific method?* ☐
2 *Define the term 'hypothesis' and explain its importance in the scientific method?* ☐
3 *Distinguish between independent, dependent and controlled variables in a first-hand investigation?* ☐

FAIR TESTING

Investigations and problem solving

REVISION TEST

1 Some hypotheses are unscientific because they cannot be tested by experiment. Say whether or not each hypothesis below is scientific and why.

a Solid rubber balls fall at the same rate as hollow rubber balls. (1 mark)

b Polyester fibres are stronger than cotton fibres. (1 mark)

c Lobster is tastier than prawns. (1 mark)

d The proportion of intelligent people living in Australia has increased considerably over the last hundred years. *Hint 1* (1 mark)

e Sight is damaged by bright, high-frequency lamps. (1 mark)

2 The following diagram shows some ideas that Marilyn jotted down when she decided to investigate the problem below.

Problem: Do small pieces of foam rubber heat up in the Sun at a faster rate than larger pieces of foam rubber?

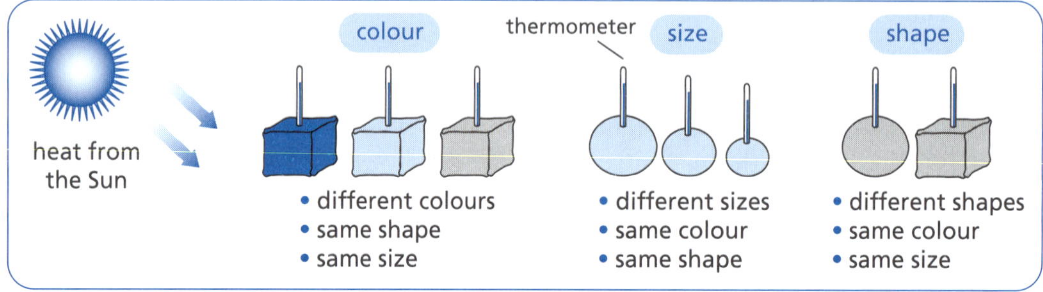

a Name the variables that Marilyn has identified. (1 mark)

b In her first set of experiments Marilyn investigated how colour affected the rate at which the rubber warmed up.

i What variables did she control? (1 mark)

ii What is her independent variable? (1 mark)

iii What is her dependent variable? (1 mark)

iv Describe the method for this experiment. (4 marks)

3 Yeast has been used for thousands of years to ferment grape sugar to make wine. The yeast microbes use the sugar as a food source. As they use the sugar, they produce carbon dioxide and ethanol (alcohol). The release of the carbon dioxide can be used to measure the speed of the fermentation process. Fred decided to investigate the rate of fermentation as a function of temperature. He had observed in some qualitative experiments that the mixture of sugar and yeast bubbled faster at 30 °C than at 20 °C, but at 40 °C there was no bubbling at all.

a Explain what is meant by a qualitative experiment. (1 mark)

b i Fred observed no gas bubbles at 40 °C. Suggest an inference that may explain this observation. (1 mark)

ii Using this inference, make a prediction that could be tested. *Hint 2* (1 mark)

iii State the expected results if the prediction was tested. (1 mark)

c Fred decided to perform a quantitative experiment to measure the change in fermentation rate as the temperature is altered. He used the apparatus shown on the next page.

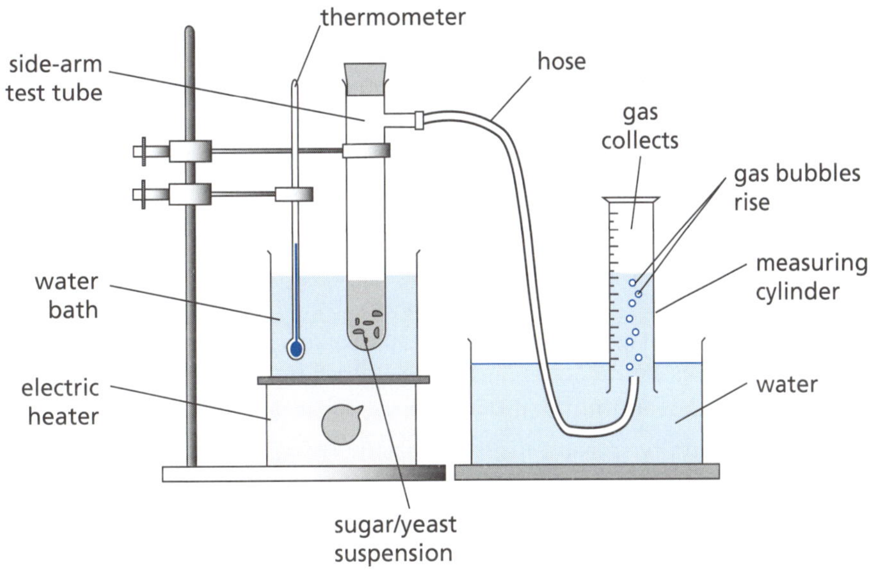

i Explain how the experiment is performed using this apparatus. *Hint 3* (4 marks)
ii State the independent variable in this experiment. (1 mark)
iii State the dependent variable in this experiment. (1 mark)
iv Identify at least three controlled variables in this experiment. (3 marks)
v This experiment is quantitative. Explain how this differs from a qualitative experiment. (1 mark)

d The results of the experiment are tabulated below.

Temperature (°C)	Gas volume (mL)
20	18
22	20
24	23
26	27
28	32
30	38

i Draw a line graph of this data. (5 marks)
ii Write a conclusion for this experiment. (1 mark)

4 True or false?
a Inferences are alternative explanations of an observation. (1 mark)
b Generalisations are true for most related observations. (1 mark)
c A hypothesis is a scientific guess. (1 mark)
d A scientific law is always true and never needs modification. (1 mark)
e In an experiment to determine the temperature of the atmosphere at increasing altitude, the altitude is the dependent variable. (1 mark)

Hint 1: Have the same standard tests been given to Australians of the same age over 100 years?
Hint 2: Yeast is a living microscopic fungus.
Hint 3: Set out your answer in a logical sequence and make sure you include uses for all equipment.

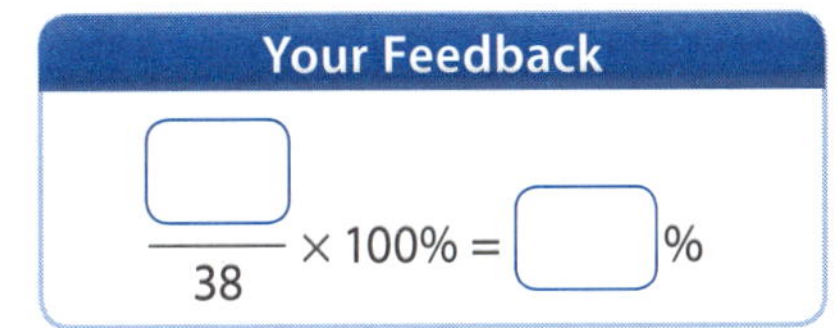

WORKING IN THE LABORATORY

Investigations and problem solving

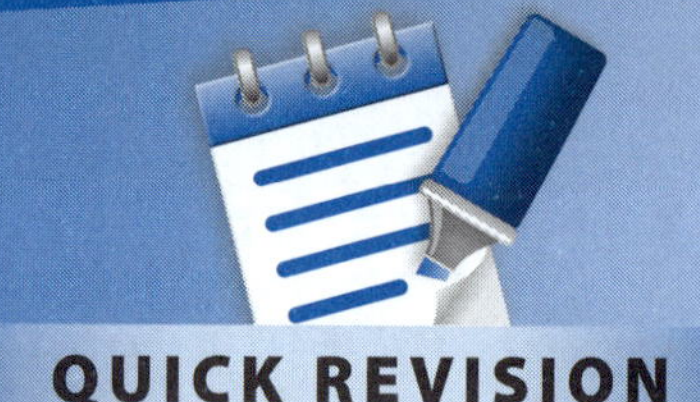

QUICK REVISION

1 Working safely in a school laboratory is a priority as this will minimise the __________ of accidents. Some rules are very obvious such as not __________ in the lab and paying attention to instructions at all times. It is important that you never bring or __________ food in a laboratory as the benches are not necessarily clean. Chemicals may not have been completely __________ up after a previous laboratory class. Never place paper scraps in the sink as this will __________ the drain. If you accidently break a test tube or beaker, be very careful in cleaning up the glass fragments. Laboratories have special glass-only bins for __________ glassware. If glass went into the paper bin then there is a danger of someone being __________.

In all first-hand investigations __________ glasses must be worn. Your eyes are easily damaged and during first-hand investigations chemicals may __________ up into unprotected eyes. Should chemicals such as dilute acids and bases splash into an eye, the eye must be treated quickly using __________ amounts of water. Eye-washing equipment should also be available.

Chemicals may accidentally splash onto your clothes. This can be avoided by wearing a __________ coat. If chemicals do splash onto your clothes, the clothing should be quickly removed and washed to avoid damage to the fibres of the cloth or to the dye. Some chemicals such as hydrogen peroxide and chlorine solutions will __________ the dyes in the cloth. Students are not allowed to use concentrated chemicals. Most solutions have been __________ to minimise risk. The following photo shows a scientist wearing a lab coat, rubber gloves and eye glasses to increase safety when working with chemicals.

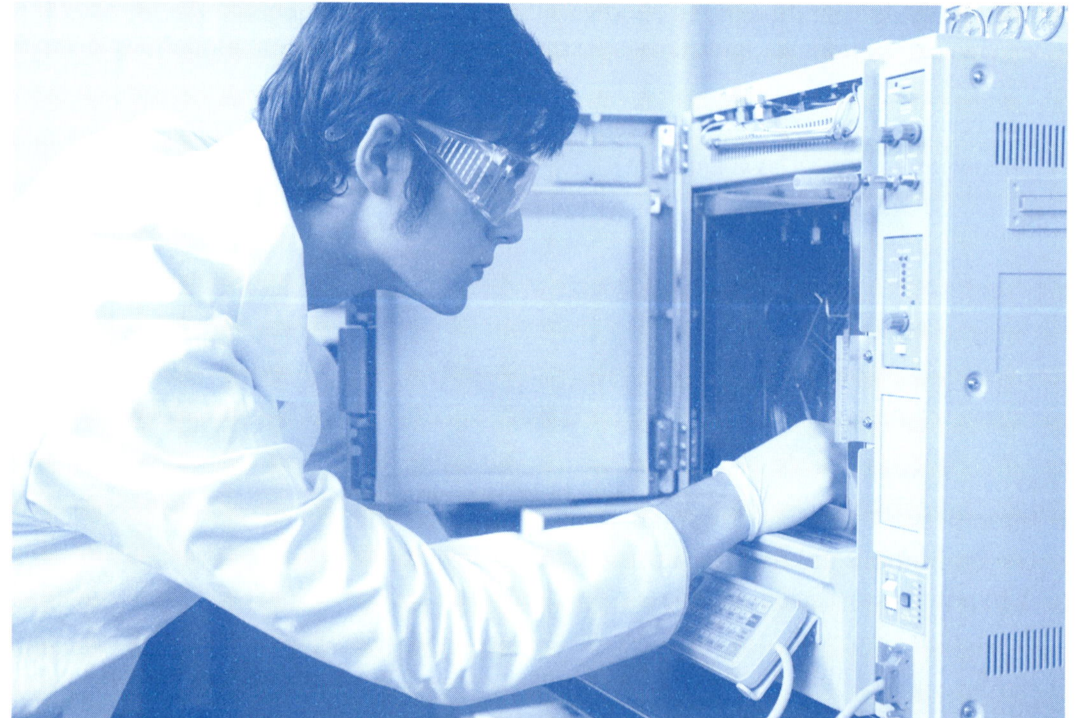

2 It is vital that you do not __________ yourself when using a Bunsen burner or a hotplate. Once heating is finished, the equipment should be left to __________ down until it is warm or cold and safe to touch. When using a Bunsen burner, you should leave it on a __________ safety flame when not in immediate use. This can be achieved by rotating the collar to __________ the air hole. The yellow flame produces __________ which would blacken the equipment if

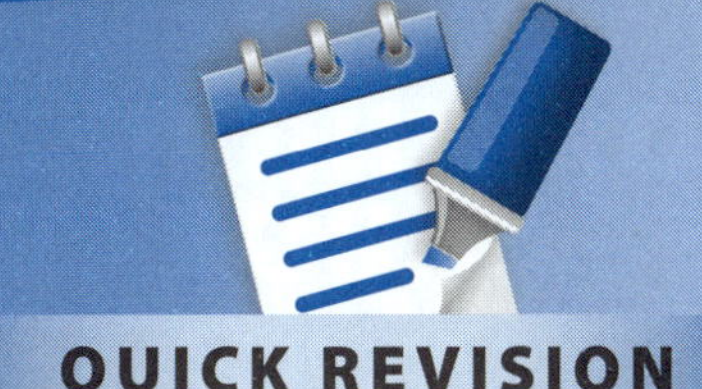

used for heating. When you are ready to heat a beaker or other glassware, rotate the collar to __________ open the hole and use the blue flame for heating. Maximum heating is achieved when the air hole is __________ open. Should an accident occur and materials catch fire, sand is available to __________ the flames. If clothing catches fire, the student should roll on the floor to help extinguish the flames and be quickly wrapped in a __________ blanket. Once the flames are extinguished, the cold running water from a shower or tap should be used to treat any __________. This treatment should continue for up to 20 minutes while __________ attention is sought. Ice is also useful for small burns to the hand.

Chemicals such as kerosene, turpentine and alcohol are __________ and so great care should be taken with their use in order to avoid fires. Never bring stock bottles of flammable liquids back to your workbench. Never leave the __________ off these fuels when a lit Bunsen burner is nearby as the vapours may catch alight. Flammable chemicals should be heated by a hot __________ bath using a hotplate to heat the water in the bath.

FLAMMABLE
LIQUID

Heating water in a test tube can also be dangerous. The tube must be held near its __________ end with a __________ test tube holder. The open end should not point at anyone and the tube should be moved in and out of the flame gently to ensure no spot is __________ too much. This process ensures that steam and hot water is not explosively ejected.

3 Live animals are sometimes used in schools for general __________. Animals such as cows and sheep may be kept and used to teach students how animals are __________ for on a farm. Small animals such as mice, birds, fish and guinea pigs may also be kept and studied. Just like domestic pets, these animals must be fed and __________ for in a humane way. In some science courses, dead animals may be used for __________ to compare their organs with ours. These animals have been purchased from a registered supplier who has authority to __________ animals for dissection. Once a dissection is complete and the results are recorded, the remains should be treated with respect and disposed of in an __________ manner. At universities and various medical institutes, live animals can be used to undertake __________. This research may be aimed at testing newly developed drugs to determine their toxicity. Once lab animal testing is complete and if no toxic effects are noted then testing on __________ volunteers can begin. Live animal research is objected to by animal __________ groups. They believe that no animal should suffer in this way.

Answers **1** risk; running; eat; cleaned; clog (block); broken; cut; safety; splash; copious (large); lab; bleach; diluted **2** burn; cool; yellow; close; soot; partly; fully (completely); extinguish; fire; burns; medical; flammable; lid; water; open; wooden; heated **3** observations; cared; cared; dissection; breed; acceptable; research; human; rights

WORKING IN THE LABORATORY

Investigations and problem solving

1 Many first-hand investigations are performed in a science laboratory and it is important you learn how to avoid **accidents**. Most of these can be avoided by using common sense, paying attention and being careful. If an accident does happen, you must inform your teacher immediately. Do not eat or drink in the laboratory. Your food can become contaminated with poisonous chemicals accidentally spilt on the desk or by fumes in the air. Always report broken glass to the teacher so that it can be safely disposed of into special glass-only bins.

It is very easy to damage your eyes by having something splash into them. Take care not to splash chemicals—pour them from one container to another carefully. Protect your eyes by using **safety glasses**. This is especially important when handling **corrosive** substances and hot liquids. If something does accidentally splash into your eyes, wash it out with plenty of water. If any substance makes contact with your body then it must be quickly removed. Often the best way to do this is to use running water or a shower if one is available in the lab. Do not sit on benches used for experiments. Your clothes need to be protected. This can be achieved using a lab coat or apron. If chemicals spill onto your clothes then they need to be removed quickly before the chemicals penetrate down to the skin.

You should become familiar with chemical safety labels like those shown below.

2 When you are using Bunsen burners, you need to be especially careful. Always light them with matches or a taper, never with burning pieces of paper. If you have long hair, tie it back out of the way. Hair can easily catch fire. Your laboratory should have a **fire blanket**, a fire bucket with sand and a **fire extinguisher**. You should know where they all are so that you can get them quickly if you need to. Mild burns can be treated by holding the burnt area under cold running water or by applying ice. If your hair or clothes catch fire, quickly drop to the floor and roll over to smother the flames. The fire blanket should be quickly wrapped around you to make sure air cannot get to the fire. Once the fire is out, call for emergency help and apply first aid Alert the teacher or other responsible adult immediately. Treat burns quickly with cold running water or iced water for at least 20 minutes. Do not put ice directly onto the burn or scald. Stay calm until help arrives.

Never touch hot apparatus. Always allow it to cool down. If you are going to heat material in a test tube, you should hold the tube near its open end with a wooden **test tube holder**. Do not use metal tongs as they slip on the glass. The open end of the test tube should point away from yourself and from other students. Never look into the open end of the tube while it is being heated. If you are going to heat a solution in a beaker, the apparatus should be set up in the middle of the bench and not near the edges. Set up the tripod, gauze and Bunsen burner but do not light the gas until all equipment is ready. When you heat the water, use a blue flame as it is not sooty and produces more heat. A yellow flame is only used as a **safety flame** when the Bunsen is not immediately in use. Never heat glassware with a yellow flame; it doesn't heat well and leaves soot on the underside of your glassware. Never place flammable chemicals anywhere near a lit Bunsen burner.

3 Animals cannot be used in first-hand investigations unless the school seeks special permission from the appropriate authorities. This occurs at universities or medical institutes where such research is essential to discover the possible toxicity of drugs as they are developed. Many **animal rights** groups are opposed to using animals for such research.

In schools you are allowed to keep live animals such as mice, guinea pigs, birds and fish but these animals can only be used for observation. They must be cared for and treated humanely. Sometimes dissections of dead animals such as rats and frogs can occur. These animals have been bred for this purpose and the companies that breed them are registered by the authorities. Your school can purchase these animals for use in the lab. When you perform a dissection, the dead animal must be treated with respect and the remains properly disposed of.

Animal parts are sometimes examined, such as dissecting kidneys, hearts or eyes. These parts can be ordered through appropriate sources such as the local butcher. They are sourced from previously killed animals and no animal is deliberately killed just for these organs. The following photo shows a frog dissection using a frog bred for that purpose.

Checklist

Can you:

1 *Explain why safety glasses and other protective clothing should be worn during first-hand investigation involving dangerous chemicals?* ☐

2 *Explain how to set up equipment involving a Bunsen burner and the ways in which burns should be treated?* ☐

3 *Explain the requirements for keeping live animals in a laboratory?* ☐

WORKING IN THE LABORATORY

Investigations and problem solving

REVISION TEST

1 The following diagram shows a student pouring some dilute hydrochloric acid from one beaker into another.

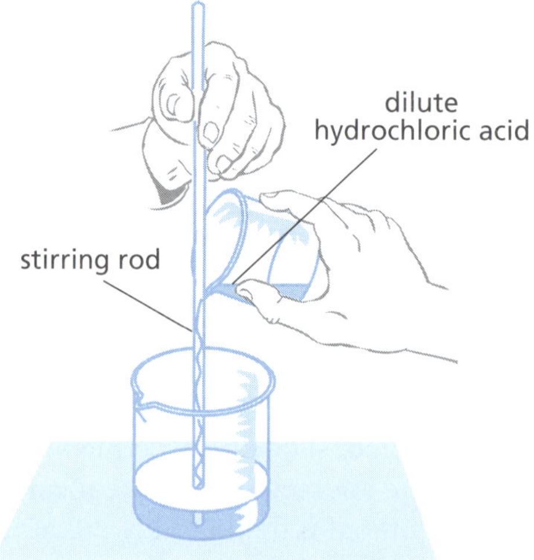

- **a** Explain why the student is using a glass stirring rod as the liquid is poured into the second beaker. (1 mark)
- **b** Explain why safety glasses are essential in this procedure. (1 mark)
- **c** A Bunsen burner with a blue flame is nearby but it is not currently in use.
  - **i** Is there a flammability safety issue in this example? Explain. *Hint 1* (2 marks)
  - **ii** Is it safe to leave the burner with a blue flame under these circumstances? Explain. (2 marks)

2 Read the following list of events that were observed in a video recording of a Year 8 laboratory experiment in which water was being heated in a beaker. For each observation, explain why the action is dangerous or a poor experimental method.

- **a** The beaker was filled to the top with water and placed on the gauze on top of the tripod. (1 mark)
- **b** The Bunsen burner was lit with a scrap of paper. (1 mark)
- **c** The collar of the Bunsen burner was turned until the hole was closed. (1 mark)
- **d** While the water was boiling, the student went over to speak to her friend in another group. (1 mark)

3 True or false?

- **a** Rats can be bred in schools and used for dissections. (1 mark)
- **b** Long hair should be tied back when conducting first-hand investigations. (1 mark)
- **c** Betty burnt her finger tips when she touched a hot tripod. The best first-aid treatment is to wrap the hand in a cloth and exclude the air. (1 mark)
- **d** Food can be consumed in a lab as long as no experiments are being conducted. (1 mark)
- **e** A gentle blue Bunsen flame can be used to heat a test tube half-full of salt water. (1 mark)

4 A Year 8 student is supplied with the following equipment: Bunsen burner; evaporating basin; gauze; tripod; beaker of salt water. The aim of the experiment is to evaporate the salt water to obtain crystals of salt.

a Draw a labelled diagram of the assembled apparatus before the Bunsen burner is lit. (5 marks)

b Discuss the safety issues involved in this experiment. (3 marks)

5 It was proposed that students make a study of the behaviour of white mice at school. Discuss the issues involved in conducting such a study. *Hint 2* (3 marks)

6 The following diagram shows the apparatus for an experiment in which hydrogen gas is being generated and collected. The reaction involves magnesium ribbon and hydrochloric acid.

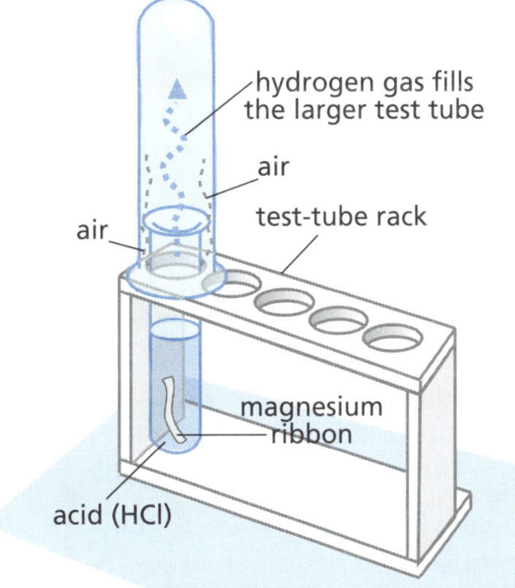

a Identify a safety issue that must be considered when using hydrochloric acid. (1 mark)

b Explain why a lit Bunsen burner should not be near this apparatus. *Hint 3* (1 mark)

Hint 1: Can you name chemicals that are flammable or non-flammable? Water solutions are normally non-flammable.

Hint 2: What things are needed to treat the mice humanely and allow them to live a normal life?

Hint 3: Is there flammable material nearby?

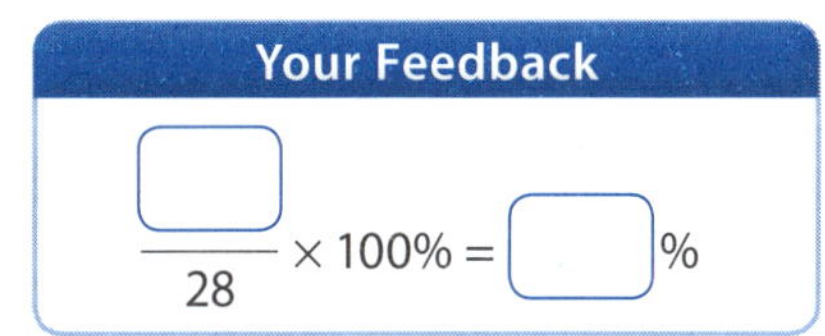

PAGE 204
PAGE 212

FIRST- AND SECOND-HAND INVESTIGATIONS

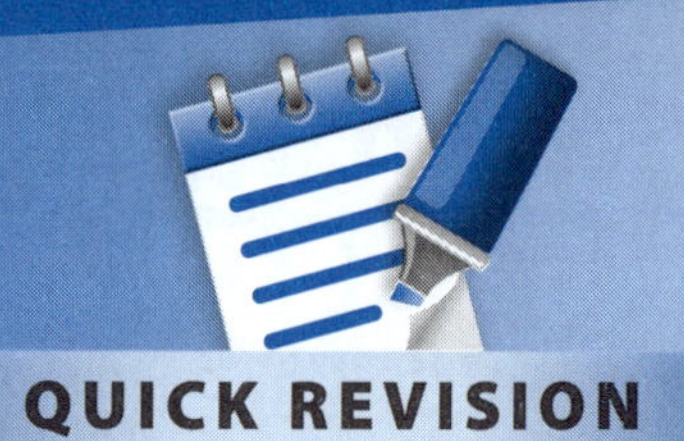

Investigations and problem solving

QUICK REVISION

1 Whether working alone or working in a __________, each has its own advantages. And people have their preferences about this, too. Some __________, and students, prefer to do research on their __________, while others like to be part of a team. Sometimes the decision can be made simpler by the complexity of the topic being researched, and at other times your teacher may nominate a lab __________ for you. But you need to be flexible enough to be able to work both ways. Some students prefer to work on their own as they are in __________ of what is happening and do not have to rely on others to come up with part of the solution to the __________. Some students may have experienced instances where their partner does not perform the necessary work, or where the work is not of a high enough __________. This puts pressure on the other team members to work harder to make up for the poor work of this person.

2 Many questions and problems in __________ today are too involved for only one person to solve. Complex and involved __________ make the process of collaboration necessary and real. There are several benefits of students or scientists working __________ in teams.

- There is increased achievement, engagement and positive __________. For instance, those working in groups can be introduced to new ideas that conflict with their own knowledge and understanding. This can lead them to seek new __________ to explain the conflict or to attempt to clarify and justify their own __________.
- Group work can produce new approaches to solving problems that __________ may not have known before working together. Individuals then adopt these approaches to use in future __________ solving.
- Individuals in groups benefit by giving and __________ help. Those giving help need to clarify and reorganise their understanding, allowing them to __________ the material better. Receiving help may fill in gaps in a person's understanding.

3 A blind experiment is often used when the subjects in the experiment are __________. For example, if testing out a new drug, some people may be given the drug to be tested while another, but similar, __________ is given a placebo (a similar-looking drug but not containing any __________ ingredient). By keeping the subjects blind about which __________ they are taking, there is less chance the results will be influenced by this knowledge. Knowing which drug is which might lead to conscious or __________ bias on their part. In a double-blind experiment, both the __________ and the person administering the test don't know which is the real drug or the __________. Only a third person who prepared the samples earlier knows. Double-blind trials are used where there is a __________ that the subjects might be influenced by their interaction with those administering the test. It could be that the experimenter has an expectation of what the __________ ought to be, and may consciously or subconsciously influence the behaviour of the subject.

4 Primary sources of information are arrived at __________-hand from the source or person. For example, a report from an experiment you performed is considered a primary source. Secondary sources are sources that are written about __________ sources. These might analyse, interpret or discuss information about the primary source such as other people's __________ results or scientific discoveries. Articles in newspapers and magazines are usually considered __________ sources.

Answers 1 group (team); scientists; own; partner; control; problem (question, investigation); standard (scratch) 2 science; problems; together; attitudes; information; position (point of view); individuals; problem; receiving; understand 3 people; group; active; drug; subconscious (unintended); subjects; placebo; risk (chance, possibility); outcome (results) 4 first; primary; research; secondary

FIRST- AND SECOND-HAND INVESTIGATIONS

Investigations and problem solving

REVISION SUMMARIES

1 Open investigations are those where students take the initiative to find answers to particular scientific problems. The problems require a level of **investigation** that yields information leading to answers. A scientific problem requires the student, or team of students, to:

- plan a route of action
- conduct the activity and collect the necessary data
- organise and **interpret** the data
- reach a **conclusion** which is communicated in some form.

Planning and problem-solving are two features that distinguish this type of investigation from other kinds of practical work. In practice, the order may not be as shown here. For instance, while carrying out the experiment, it may be discovered that more planning is necessary.

Working individually requires students (and scientists) to:

- take **personal responsibility** in the planning and performance of the task
- realise there are goals and timelines that need to be adhered to
- **persevere** with the task to arrive at a satisfactory end point
- accept personal responsibility that the working environment is safe for the individual and for others around
- evaluate the **effectiveness** of the work in completing the task.

2 Working **collaboratively** in teams requires students (and scientists) to:

- identify that each team member has different roles to play
- negotiate and **assign responsibility** to each member of the team
- accept and perform the roles assigned by the team as a whole
- collaborate with others in setting timelines and working to an agreed set of goals
- accept **personal responsibility** that the working environment is safe for the individual and for other team members
- evaluate the process used by the team and the effectiveness of the team in finishing the task.

Working in a team can help to generate ideas as you discuss them, refine these ideas and add **different viewpoints** to the mix. Solving problems and dealing with conflicts can also be instructive for the entire team. Working as a team can help with social interaction (you need to get along with other members) and it also provides support when things get difficult.

But group work does have drawbacks as well. Some students find it difficult to work together and require good models and training for the process. The standing of individuals within a group makes some students want to lead, while other students are happy to be followers. The person whose ideas are respected in general may not be the person with the best understanding of the problem to be solved.

3 **Blind experiments** are important tools in many fields of research such as medicine, psychology and the social sciences and in natural sciences such as physics and biology. Blind experiments are crucial in testing the value of new medicines and painkillers.

Blind experiments are used when it is vital that the subjects do not know (are 'blind to') which treatment they are getting. That way they can be completely honest about the effects of their treatment. Here, researchers often give one group of subjects a fake or placebo pill, which does not contain any of the drugs being tested. So the subjects do not know which they are getting: the active medication or the placebo. Even so, a difficulty sometimes arises because the subjects

(cont.)

in a blind experiment could convince themselves that they are getting the active medicine, and this can cloud the results.

In order to remove any **bias** from the investigator, a **double-blind experiment** is used. In this a third person prepares tablets in two batches; one containing the new treatment and the other one a placebo. A number of subjects are asked to volunteer to take part in the experiment, and they are randomly assigned to the experimental or to the control group. The subjects are blind to which group they will be in. And this time the investigator does not know either; so the investigator is also blind to the treatments. Eventually the effectiveness of the treatments is analysed and a report summarises the results. Only then does the third person reveal which pills were taken by which group.

4 A **primary source** is an original object or document, that is, the raw material or first-hand information. Primary sources include results of your own experiments, statistical data or where a direct observation was made.

A **secondary source** is something written about a primary source. Secondary sources could be comments on, interpretations of or discussions about the original material. This is second-hand information. Secondary source materials include articles in newspapers or popular magazines, books and journals. These may talk about or appraise someone else's original investigations. Most of the information we obtain and use in science is from secondary sources. Secondary data is also used to gain preliminary insight into some research problem we are planning to investigate.

Advantages of using secondary sources include the following.

- The research process can be completed quickly.
- You only need to locate the source of the data and extract the required information.
- Secondary research is generally less expensive than primary research.

But there are also some disadvantages of using secondary data, including the following.

- The secondary information you are looking for may be either not available or is only obtainable to an inadequate extent.
- You can't always rely on its accuracy or reliability.
- Data may not be in the format or use appropriate units you require.
- Much secondary data is dated and may not reflect the results of more recent research.

As a general rule, secondary data should be examined thoroughly prior to conducting primary research. This will provide a helpful background and help to recognise key problems and issues that should be addressed by the primary investigation.

Checklist

Can you:

1 *Recognise the advantages and disadvantages of working individually?* ☐
2 *Recognise the advantages and disadvantages of working collaboratively as a group?* ☐
3 *Describe blind and double-blind experiments?* ☐
4 *Recognise the advantages and disadvantages of secondary source research?* ☐

FIRST- AND SECOND-HAND INVESTIGATIONS

Investigations and problem solving

30 MINUTES

REVISION TEST

1 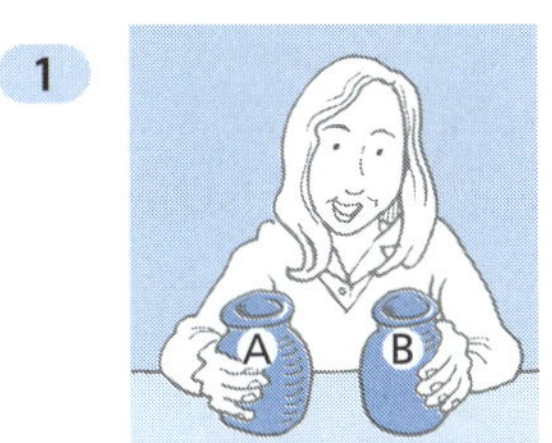

Mary is experimenting with making chocolate biscuits, but she can't decide which is better: making them with cooking chocolate or with cocoa. So she makes two batches, placing one batch in a jar marked A and the other batch in a jar labelled B. She asks some friends over for a party and asks them to try each kind of biscuit and tell her which they prefer.

a Is this a single-blind or double-blind test? How do you know? (3 marks)

b In order to obtain fair results, what are some of the things Mary must keep the same while making these biscuits? (2 marks)

c Would it be fair if she only asked one or two friends over? Explain. *Hint 1* (2 marks)

2 At police stations witnesses or crime victims are often shown a group of photos to pick out the suspect. There is a growing movement by police to shift to a double-blind process where the officer showing the photos does not know which photo is of the suspect. Why is this an advantage? (2 marks)

3 The following diagram shows the steps involved during a double-blind experiment for patients showing a particular medical condition to test the effectiveness of a new experimental drug.

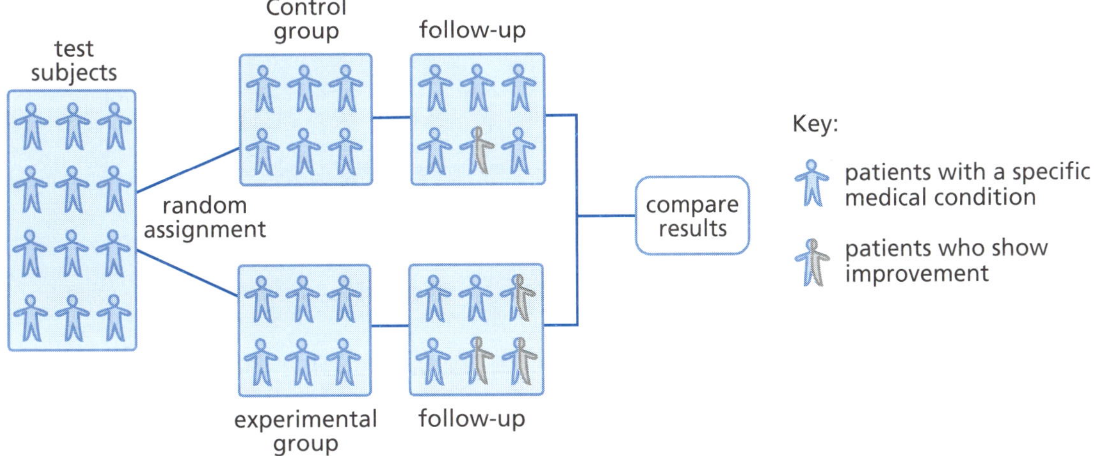

a Describe the steps involved in a double-blind experiment. (4 marks)

b Why is it important that a number of patients be assigned to each group? (2 marks)

c What are some features the control group and experimental groups should have in common? *Hint 2* (2 marks)

d Why shouldn't the test subjects know which drug they are taking? (2 marks)

e Why shouldn't the doctor administering the drug know which drug is which in a double-blind experiment? (3 marks)

f Can you think of any ethical issues that might need to be considered? *Hint 3* (1 mark)

4 a Explain why secondary data is less expensive and easier to obtain than primary data. (3 marks)

b Explain why secondary data may be outdated and not reflect the current knowledge. (4 marks)

c Explain why secondary data cannot always be relied on for its accuracy or reliability. (2 marks)

(cont.)

5 The strength of a team is the diversity of knowledge, skills and abilities that different individuals bring.
Comment on this statement. (2 marks)

6 The following photo shows a team of scientists in the laboratory. Suggest four reasons why it is important to learn to work in teams. (4 marks)

7 For each of the following, state whether they are primary or secondary sources.

a an eyewitness account of an event or circumstance (1 mark)

b interprets, analyses or explains a scientific event (1 mark)

c while attempting to be objective and impartial, you may reflect the prejudice of the investigator (1 mark)

d original documents, diaries, manuscripts or letters (1 mark)

Hint 1: Would it be better to have the view of only one or two people or of a number?
Hint 2: This is true of any controlled scientific experiment.
Hint 3: Ethical issues involve right and wrong, or what is considered good or improper in a society.

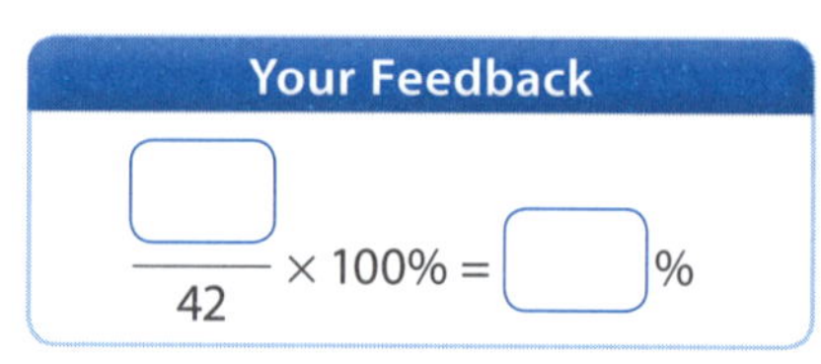

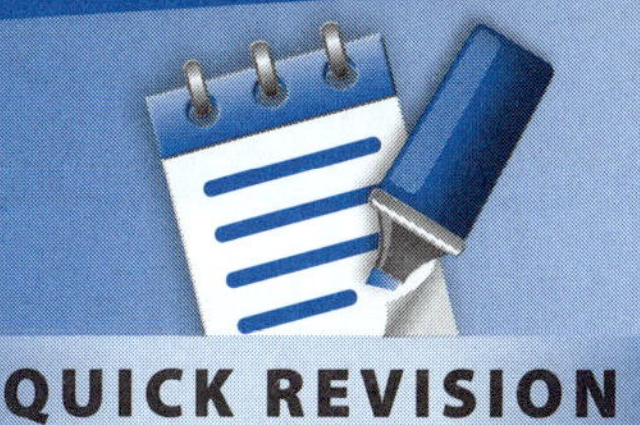

EXPERIMENTAL REPORTS

Investigations and problem solving

QUICK REVISION

1 Experimental reports are a vital part of all ________ sessions and is the main means of documenting what was done. This is where you explain what you did in the ________, what you learned and what the results meant. There is a standard format for writing these reports, although there can be some variation.

- **Title.** This is a word or phrase ________, and includes the date when the research was conducted.
- **Aim.** It should be brief and describes the main point or ________ of the experiment or investigation.
- **Background information.** Sometimes it may be necessary to inform the reader what led up to this experiment being performed or citing some secondary data that would be of ________ in performing this investigation.
- ________. List everything needed to complete your experiment, but there is no need to mention common pieces of equipment which would be ordinarily found in a laboratory.
- **Method.** This describes the steps, in ________ form, of the procedure you completed during the investigation. It needs to be sufficiently detailed so that anyone else reading this can ________ the experiment. It is often helpful to provide a ________ showing the experimental setup.
- **Results.** Numerical data are often displayed in a ________ and the results describe in words what the data means. Sometimes results can be joined with the discussion.
- **Discussion and analysis.** Any calculations based on numerical results are performed here. This section is also where you ________ the data and determine whether or not a hypothesis was accepted.
- **Conclusion.** This is often a single ________ that summarises what happened in the experiment, and whether the hypothesis was accepted or rejected.

2 What you find out by performing the experiment might be ________ (perhaps you generated a set of numbers by measuring variables such as change of weight, increase in height) or descriptive (some observation of colour change, whether or not a reaction occurred). Presenting collected numerical values in a ________ and then drawing appropriate ________ to show trends or other features is useful. If the collected information is a set of observations, write these down as you see them. In the discussion section of the experimental report, you can link the results to what you read in the literature or other ________-hand sources. Do the results you obtained support what you read or what you were told? Discuss whether or not the results supported your ________. If they did not, discuss why not. Suggest ________ that may have affected the experimental design. Discuss how they can be eliminated in the future. Discuss the possibility of using a different method or design of the experiment to ________ it, if appropriate.

3 Writing a conclusion involves ________ up the report and giving a very brief description of the results, although you should not go into too much detail about this. Anybody reading the conclusion has possibly already read the entire report, so the ________ merely acts as an aid to memory. Always use the ________ tense when writing the discussion or conclusion section of a practical report.

Answers **1** laboratory (experimental, practical); experiment; heading; purpose; relevance (importance); **Materials**; point (dot); duplicate (repeat); diagram; table; interpret (discuss); paragraph (sentence) **2** numerical; table; graphs (charts); second; hypothesis; biases (problems, issues); improve **3** summing; conclusion; past

EXPERIMENTAL REPORTS

Investigations and problem solving

REVISION SUMMARIES

1 The following is an example of a **laboratory report** for an experiment. Steps need to be clear so that anyone else reading this can follow the procedure and repeat the experiment.

Measuring transpiration in plants

A clear **title** uses few words.

Aim

To measure the rate of water uptake by a plant shoot using a potometer.

The **aim** describes the purpose for the experiment.

Background information

A potometer is a device used to measure the rate of water uptake of a leafy shoot. While there are several designs, it basically consists of a sealed vessel of water with a plant cutting inserted in such a way that moisture can escape only by absorption and transpiration. Water is taken up by the plant because of photosynthesis and transpiration. For it to work properly, the potometer must be assembled under water, so a large sink is necessary. Transpiration is the process by which water absorbed by plants through its roots is evaporated into the atmosphere from the plant surface, such as from leaf pores.

The shoots also need to be cut under water. If air gets into the xylem vessels then air locks can form preventing the experiment from working properly.

Some people may not have heard of a potometer, how it works and what it does, so some background information is useful.

Safety issues

Some people may find the sap from plants irritating. Also take care when cutting the plant shoot and when working with glass to prevent accidents.

Sometimes **safety** issues need to be considered.

Materials required

- *a plant cutting, preferably with thin waxy cuticles*
- *scissors or scalpel*
- *a calibrated pipette to measure water loss*
- *a length of clear plastic tubing*
- *wax or chewing gum to form an airtight seal between the plant and the water-filled tubing*
- *large clear polythene bag (for humidity experiment)*
- *fan*
- *desk light*
- *rectangular glass tank with water.*

Materials listed should only be those which would need to be brought into the laboratory and are not readily available.

Method

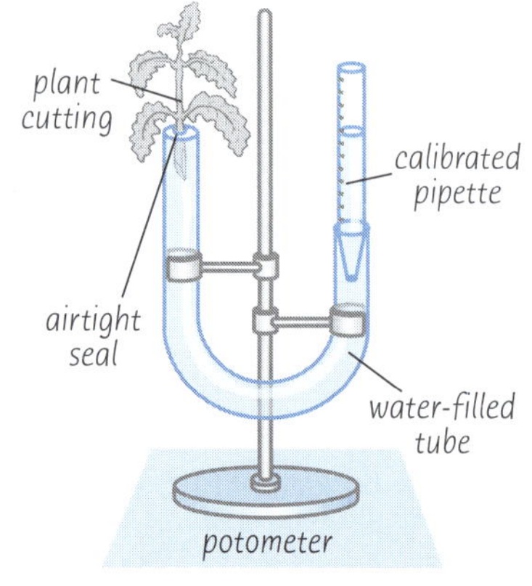

1. Set up the potometer as shown in the diagram. The apparatus needs to be assembled under water.

A **diagram** is very useful in showing how the apparatus should be assembled.

2. Allow the leaves to dry thoroughly before beginning the experiment.

3. Perform the experiment at 45-minute intervals under the following conditions: ordinary room (control), humidity, wind, bright light.

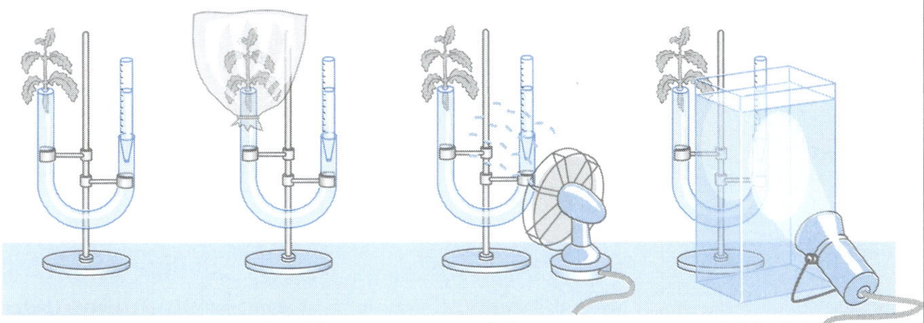

4. Make measurements of how much water is taken up every 4 minutes for each of these four environmental conditions, by reading the calibrated pipette.

5. Allow at least 10 minutes before repeating the experiment for the plant cutting to acclimatise to its new environment.

Steps in the **procedure** can be numbered.

Notice that the first person (I, me) and second person (you) are rarely used when writing experimental reports.

2 **Results** from the experiment need to be recorded. These could be numerical data generated or a description of what was observed. Diagrams can also be included.

Results

	Total transpiration (mL)			
Time (min)	Control	Humid	Wind	Bright light
0	0	0	0	0
4	1	1	2	2
8	2	2	3	4
12	4	2	5	6
16	4.5	3	8	7
20	6	4	10	10
24	8	4	12	11
28	10	5	14	13
32	11	6	15	15
36	12	6	17	16
40	14	6	18	19
44	16	6	20	21

In this experiment the results were numerical and are suited to being reported in a **table** (tabulated). In other practicals results might be a description of what was observed and would be written in sentence form.

(cont.)

EXPERIMENTAL REPORTS *(continued)*

Investigations and problem solving

Discussion and analysis

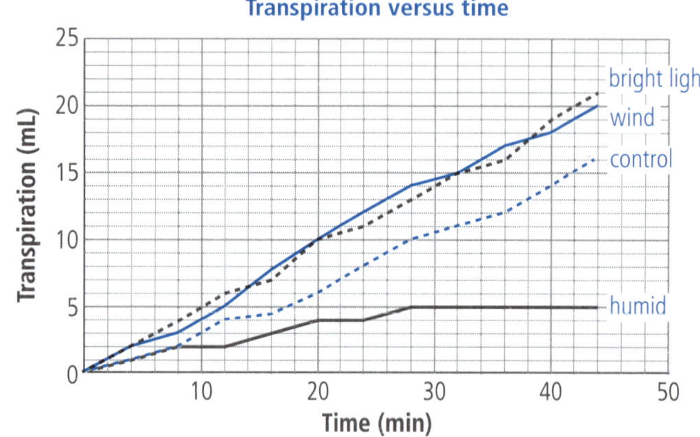

The least amount of transpiration occurred under humid conditions. This was modelled in the experiment by using a polythene bag around the leaves to retain moisture. Initially, transpiration was similar to that in the control but, as the bag contents became more humid, the rate of transpiration decreased. Bright light and wind showed similar results to each other, and more transpiration than the control. In the case with the bright light, the light was passed through a rectangular water bath to capture any heat and allow only light through. The results are in accordance with second-hand reports. Wind removes any moisture from near the leaf, hastening more water to transpire. Light has the effect of increasing the rate of photosynthesis thus using more water.

In the **discussion** and **analysis** section you make sense of these results, answer any questions that might be posed and discuss the results of the experiment in the light of second-hand research.

A **line graph** for these continuous data is the preferred method of graphical display.

The discussion section is also where you can discuss any shortcomings of the experiments and make suggestions as to how it could be improved.

3 A suitable **conclusion** needs to be written which answers the problem posed in the aim. The conclusion neatly sums up what you learned from the experiment.

Conclusion
A potometer is a suitable device to show the uptake and transpiration of water. Varying the environmental conditions of the plant affects the rate of uptake of water.

A conclusion should only be a sentence or two in length.

Checklist

Can you:

1 *Describe what a practical experimental report should consist of?* ☐
2 *Discuss the results of an experiment in the light of second-hand research?* ☐
3 *Write a conclusion for a first-hand investigation?* ☐

EXPERIMENTAL REPORTS

Investigations and problem solving

REVISION TEST

1 In designing a new rocket fuel mix a scientist mixed two fuels, P and Q, together and measured the height a rocket reached. The results are given in the table below.

Trial number	Mass of fuel P (kg)	Mass of fuel Q (kg)	Altitude reached (m)
1	8	10	800
2	10	10	1000
3	12	10	1200
4	14	10	1350
5	16	10	1300
6	18	10	1200
7	20	10	1100

a Which is the:

i independent variable? (1 mark)

ii dependent variable? (1 mark)

iii controlled variable? (1 mark)

b If the scientist wishes to find the ideal mixture that makes the rocket go the highest, what might the next step in the experiment be? *Hint 1* (1 mark)

c Why is it important that the same rocket be used in each of these trials? (1 mark)

d It is essential that you only have one independent variable at a time. Why is this? (1 mark)

2 A student wanted to study the effect of a nitrate fertiliser on plant growth. She took two similar plants, planted them in similar pots of soil and set them on a window sill for a fortnight observation period. She watered each plant the same amount, but she gave one of the plants a small measure of fertiliser with each watering. She collected data by determining the height that each plant grew. State one significant fault in this experimental set-up. *Hint 2* (2 marks)

3 Estuaries are bodies of salt water that are fed by freshwater sources. Here freshwater mixes with salt water and salinity (the saltiness of the water) can vary. *Elodea* is a common aquatic plant found in estuaries where freshwater is plentiful. Patrice wanted to find out how varying concentrations of salt affect the plant *Elodea*. She set up five test tubes with *Elodea* as shown.

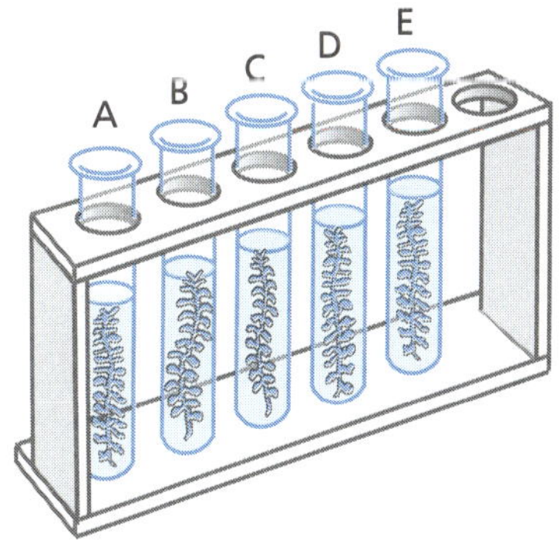

a What would be the aim of the experiment? (1 mark)

b Why should one of the test tubes contain plain pond water from where the plant came from? (2 marks)

c What solutions should she place in each of the other four test tubes? Suggest examples. (2 marks)

d She observed the plants each day for a week. What should she be looking for? *Hint 3* (1 mark)

e How would you suggest she record her results? (1 mark)

4 In another experiment with *Elodea*, Demeter set up the following apparatus to determine the rate at which oxygen bubbles are formed when the light source is placed at varying distances from the plant.

a Write an aim for this experiment. (1 mark)

b What is the reason for placing the light source at varying distances from the plant? (1 mark)

(cont.)

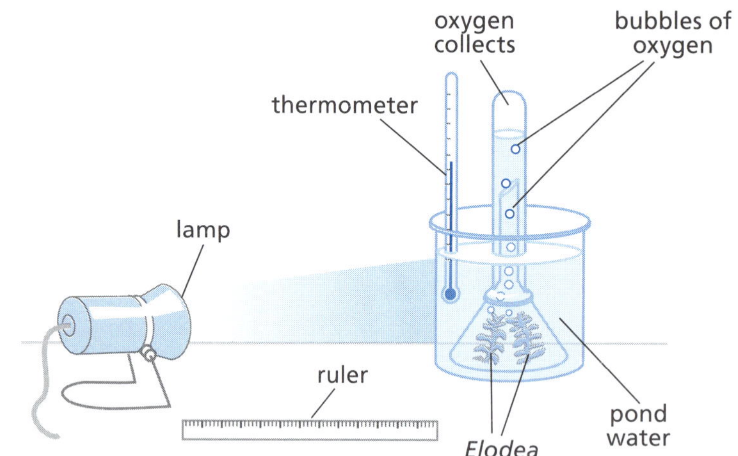

c Describe the experimental set-up for this investigation. (5 marks)

d What is the purpose of the thermometer? (2 marks)

The following graph shows the results of Demeter's experiment.

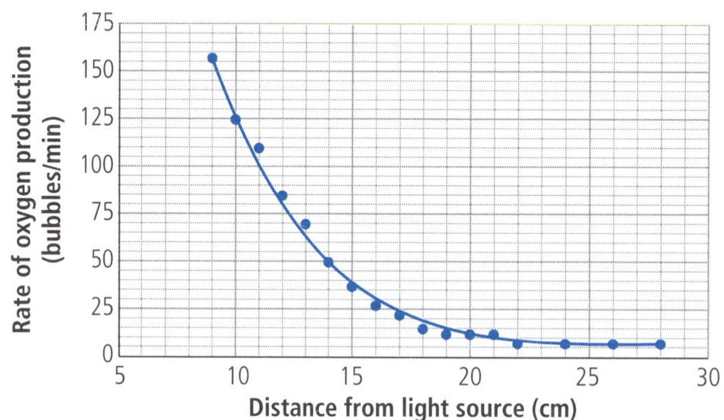

e Give the graph a title. (1 mark)

f How does the distance from the light source affect the number of bubbles produced? (1 mark)

g State two other factors that could affect the rate of bubble production. (2 marks)

h Suggest one way this experiment could be improved. Why do you make this suggestion? *Hint 4* (4 marks)

Hint 1: Remember: the scientist wishes to determine the optimum fuel mix to achieve maximum height.

Hint 2: The experiment needs to be repeated a number of times to overcome any peculiarities of any one particular plant.

Hint 3: Remember: she is trying to find an answer to the aim.

Hint 4: There are better ways to perform this experiment and obtain the answers she seeks. Can you think of one of these? Counting so many bubbles can be problematic.

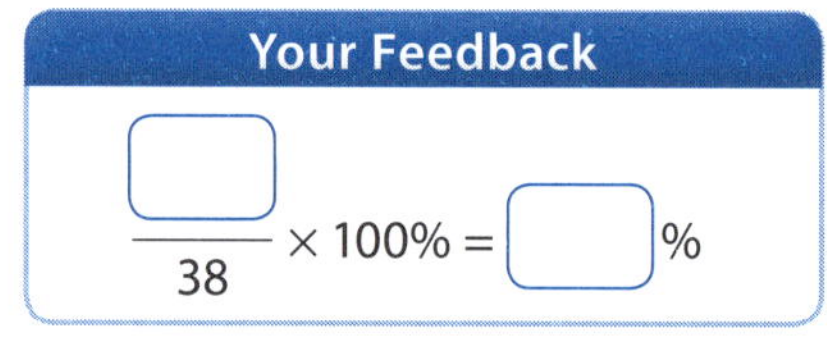

TIPS FOR THE SAMPLE EXAM PAPERS

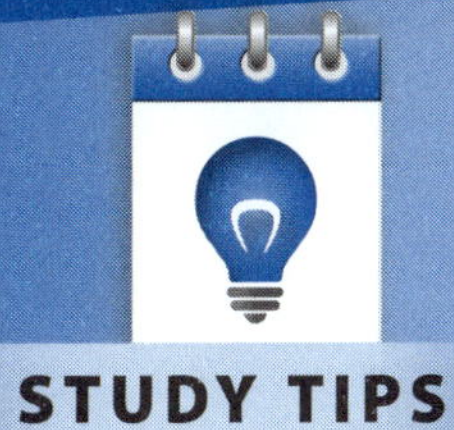

STUDY TIPS

Scope of work covered in the Sample Exam Papers

- There are two Sample Exam Papers in the next section.
- The first one covers content from chapters in the first half of the book.
- The second one covers the whole book content but with greater emphasis on the chapters in the second half of the book.

Your school's science program

- Your school may teach the content in a different order to that presented in this book.
- You will need to identify those questions that are relevant to the content you have studied in class as you prepare for major examinations.

Suggested times

- The suggested time for each Sample Exam Paper is 50 minutes.
- Keep track of time when you do each exam.
- If you have not completed the exam in that time, continue on and record the time you have taken.
- While finishing within the time given is important, it is equally important to answer the questions correctly.

Part A: Multiple-choice questions

- Part A of each Sample Exam Paper has 15 multiple-choice questions. Read the stem of the question carefully to check what is being asked.
- If a diagram is supplied, check the labels carefully as they may give important clues.
- Choose the best response. If you are unsure of the correct answer, make the best logical choice that you can.
- No marks are deducted for incorrect answers, so don't leave any unanswered. Sometimes you can eliminate a wrong alternative, improving your chance of deciding on the correct answer.

Part B: Restricted-response questions

- Part B contains 10 restricted-response questions where one-word answers are required to fill the missing space in a sentence. These questions generally test basic recall of knowledge.
- Often the answer is a key word that forms part of the definition or point being examined.

Part C: Written knowledge and skill questions

- Part C contains written knowledge and skill questions that test your deeper understanding of the topics and your ability to process new data. These exams often test your mathematical and graphing skills, as well as your ability to interpret or draw scientific diagrams and analyse information.
  - Use the mark value of each question as a guide to the length of the response required.
  - Look for key words in each question.
  - Do not restate the question in your answer.
  - You should always present your answer in a logical order.
  - If you draw a labelled diagram as part of your answer, make sure you refer to the diagram in any explanation. Always use a pencil and ruler for diagrams.
- Sometimes you will be asked a question that involves using common knowledge and applying thinking ability to a novel scientific situation.

Preparation for the Sample Exam Papers

- Study each section of the relevant chapters again as you prepare to do the Sample Exam Papers.
- It is important to revise throughout the year. As the school year progresses, go back and redo the Revision Tests and the Sample Exam Papers. This will help you revise for major school exams. Your results will show where more revision is needed.

SAMPLE EXAM PAPER

50 MINUTES

PAPER 1

PART A *Multiple-choice questions* 15 marks

1 The jelly-like substance inside every cell is called
A a vacuole.
B cytoplasm.
C a ribosome.
D a nucleus. (1 mark)

2 Digestion begins in the
A mouth.
B stomach.
C small intestine.
D large intestine. (1 mark)

3 The system responsible for generating electrical impulses allowing us to move our muscles is the
A cardiovascular system.
B muscular system.
C skeletal system.
D nervous system. (1 mark)

4 For this diagram, which statement is true?

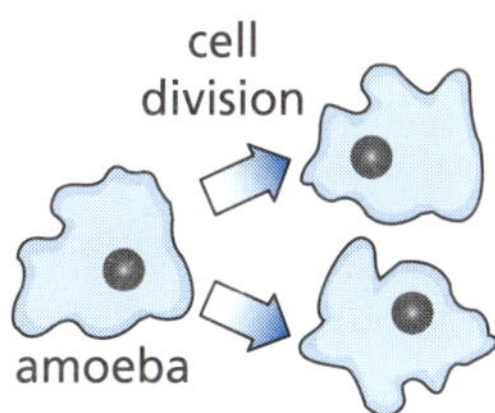

A The resulting two cells will each contain half the number of chromosomes.
B The resulting two cells will have identical DNA.
C Each of the daughter cells will take on unique personalities.
D This form of cell division is called meiosis. (1 mark)

5 DNA replication inside a cell takes place in the
A vacuole.
B cytoplasm.
C nucleus.
D ribosome. (1 mark)

6 Name the odd one out.
A oesophagus
B ureter
C appendix
D duodenum (1 mark)

(cont.)

7 Consider the following three statements.

i In solids the particles are packed the closest together.
ii In gases the particles are spread out the furthest.
iii In liquids the particles are packed fairly close together.

How many of these statements are correct?

A none
B one
C two
D three (1 mark)

8 When a sperm cell fuses with an egg cell this is called

A intercourse.
B fertilisation.
C pollination.
D reproduction. (1 mark)

9 Which of the following is true about all molecules?

A They have the same mass.
B They are made up of the same number of atoms.
C They are the same size.
D They are constantly moving. (1 mark)

10 Air enters the lungs through the

A oesophagus.
B trachea.
C alveoli.
D pulmonary artery. (1 mark)

11 Comparing the neutral atom Ca (calcium) with the positive ion Ca^{2+}, the positive ion has

A more protons than the neutral atom.
B the same number of electrons as the neutral atom.
C fewer electrons than the neutral atom.
D a different number of neutrons from the neutral atom. (1 mark)

12 The part of the blood carrying minerals, vitamins, sugars and other foods to the body's cells is/are the

A platelets.
B red blood cells.
C white blood cells.
D plasma. (1 mark)

13 Egg cells are larger than sperm cells because egg cells

A contain more cells.
B need a food store to help growth after fertilisation.
C have larger nuclei.
D have thicker cell membranes. (1 mark)

14 Many people are allergic to the droppings of dust mites. These are tiny animals related to
- **A** spiders.
- **B** wasps.
- **C** flies.
- **D** insects. (1 mark)

15 Bile is made in the
- **A** liver.
- **B** gall bladder.
- **C** stomach.
- **D** small intestine. (1 mark)

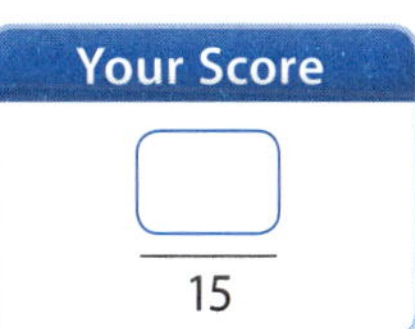

PART B *Restricted-response questions* 10 marks

Insert the missing word to complete each sentence.

16 The organelle most important in providing energy to the cell is the __________. (1 mark)

17 The blood vessels that carry blood away from the heart are called __________. (1 mark)

18 The following diagram shows two different species of organisms in the same kingdom. Both of these organisms reproduce asexually by __________ formation. (1 mark)

19 One major feature all plant cells have that animal cells lack is a __________. (1 mark)

20 White blood cells, also called __________, are cells of the immune system involved in defending the body against both infectious disease and foreign materials. (1 mark)

21 A __________ must infect a living host to reproduce. (1 mark)

22 The particles of matter shown in this diagram are in the __________ state. (1 mark)

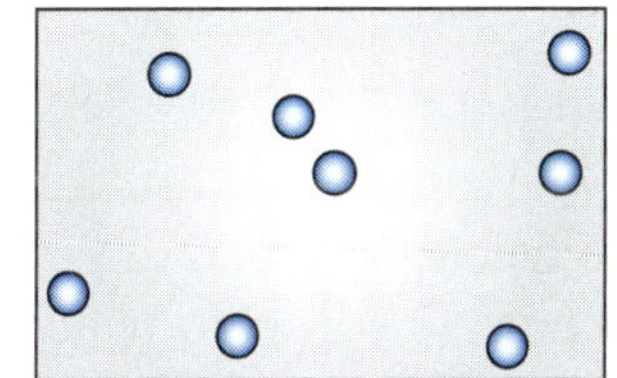

(cont.)

23 When they reach the stomach, crushed and mashed-up particles are mixed with ____________ juices. (1 mark)

24 Penicillin is an antibiotic that weakens the cell walls of ____________. (1 mark)

25 The part of the female reproductive system producing eggs is the ____________. (1 mark)

Your Score

/10

PAGE 209

PART C *Written knowledge and skill questions* 25 marks

26 Centuries ago, when sperm cells (then called 'animalcules') were first discovered using early microscopes, a few observers thought that these contained miniature humans (*homunculus* is Latin for 'little man') which grew in the womb to become the new child. This gave rise to the homunculus theory. Up until the end of the 18th century, many argued that children drew their fundamental features from the father. The mother, it was thought, provided only a womb and the raw material.

How does the homunculus theory differ from what we know today about reproduction? (3 marks)

27 a Which of the following is not an example of a chemical change? (1 mark)

- combustion (burning) of petrol
- cooking an egg
- rusting of an iron pan
- melting an ice cube
- mixing hydrochloric acid and sodium hydroxide to form salt and water

b What is the difference between a chemical and a physical change? (1 mark)

c Give two observations or clues that suggest a chemical reaction has taken place. (2 marks)

28 Write true or false for the following statements about the microscope.

a A wet mount is a freshly prepared slide where a drop of water has been added. (1 mark)

b The total magnification produced by a microscope, using a 10× ocular (eyepiece) lens and a 10× objective lens, is 20×. (1 mark)

c The iris diaphragm lever regulates the amount of light passing through the slide specimen on the microscope stage. (1 mark)

29 a What kind of bond is formed when atoms share valence electrons? (1 mark)

b What is the difference between an ionic and a covalent bond? (2 marks)

c What is the main kind of chemical bond in plastics? (1 mark)

30 a An element is a typical metal. State two properties which this element would have. (2 marks)

b Complete the following word equations.

i zinc + oxygen → ____________ (1 mark)

ii magnesium + hydrochloric acid → ____________ + hydrogen (1 mark)

iii aluminium oxide → aluminium + ____________ (1 mark)

31 Cisplatin is one of the most powerful and broadly used anticancer drugs.

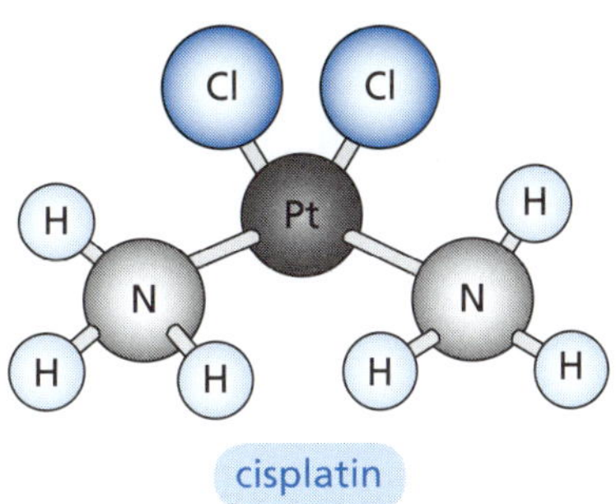

cisplatin

a Name the metal in cisplatin. (1 mark)

b How many atoms make up one molecule of cisplatin? (1 mark)

c Why is this diagram often referred to as a ball-and-stick model? (1 mark)

32 The following diagram shows parts of a flower. Use the words in the diagram to explain how flowering plants reproduce. (3 marks)

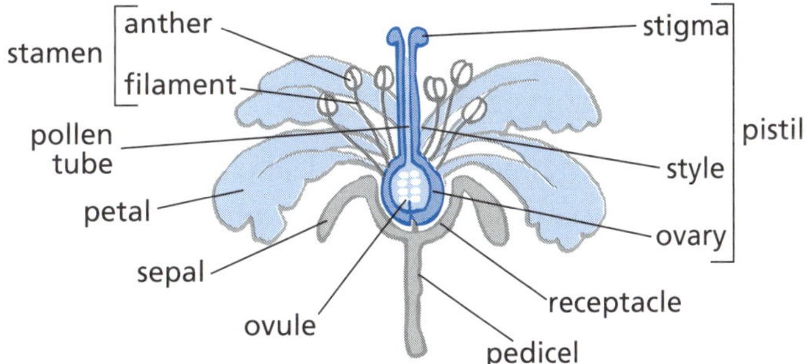

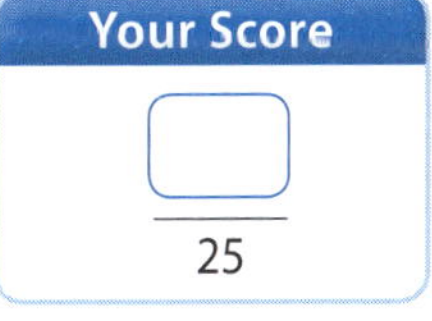

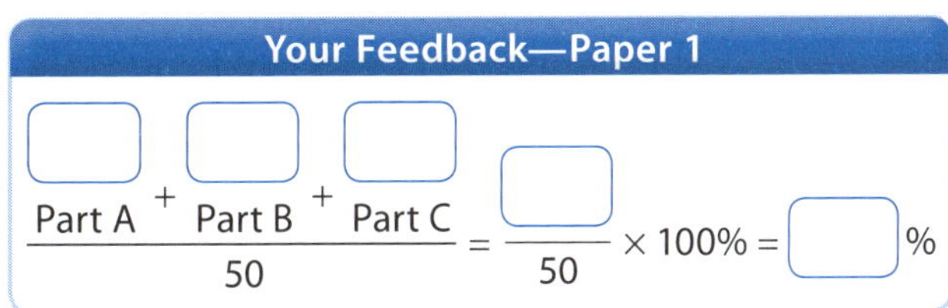

SAMPLE EXAM PAPER

PAPER 2

PART A *Multiple-choice questions* 15 marks

1 The nosepiece in a simple light microscope is
- **A** a knob that rotates to allow different objective lenses to be selected.
- **B** a knob used to raise/lower the body tube to bring the image into approximate focus.
- **C** the location where the specimen is placed on a glass slide.
- **D** an adjustable diaphragm used to alter the amount of light on the specimen. (1 mark)

2 Select the statement that is true about arteries and veins.
- **A** Oxygenated blood leaves the heart via the aorta, which is the major artery of the body.
- **B** Veins have much thicker and more elastic, muscular walls compared to arteries.
- **C** The blood pressure is higher in veins than in arteries.
- **D** Gaseous exchange occurs in the fine blood vessels called veins. (1 mark)

3 Select the true statement about ethanol (C_2H_5OH).
- **A** Ethanol is an ionic compound.
- **B** One molecule of ethanol contains six hydrogen atoms.
- **C** Ethanol is not a pure substance as it is composed of three different elements.
- **D** The ratio of carbon to hydrogen is 2:5. (1 mark)

4 Select the true statement about chemical and physical changes.
- **A** Physical changes involve new substances with new physical and chemical properties being formed.
- **B** Physical changes are quite difficult to reverse.
- **C** The combustion of candle wax is an example of a chemical change.
- **D** The condensation of water vapour is an example of a chemical change. (1 mark)

5 An example of a native metal (metal found in uncombined form in nature) is
- **A** zinc.
- **B** sodium.
- **C** platinum.
- **D** tin. (1 mark)

6 The process of converting copper ores into copper metal is referred to as
- **A** froth flotation.
- **B** electrolysis.
- **C** flocculation.
- **D** smelting. (1 mark)

7 In sustainable agriculture, companion planting and green manure is used to
- **A** keep the land perpetually fertile and to prevent topsoil loss.
- **B** reduce the build-up of diseases and pests.
- **C** avoid irreversible changes to the land, such as erosion.
- **D** reduce fire intensity during the dry seasons. (1 mark)

8 Wetlands are important because they
A provide breeding grounds for wildlife, particularly kangaroos and koalas.
B support a wide variety of plants and animals, forming different habitats and ecosystems.
C can be used to store water before it is converted into drinking water.
D are important refuges for wildlife such as wombats and possums, especially in times of floods.
(1 mark)

9 Do not sit on laboratory benches used for experiments because
A there will be little room left to carry out experiments.
B your clothes may absorb chemicals that have not been properly cleaned up from a previous experiment.
C you may accidentally turn on the power switches.
D other students behind you will have trouble seeing the teacher at the front. (1 mark)

10 Each laboratory should have a fire blanket. These are used for
A throwing over a kerosene fire on the bench.
B protection during a lab fire.
C smothering the flames when clothing catches alight.
D providing an absorbent material that will soak up the burning fuel. (1 mark)

11 The Sun produces radiant energy. The process that produces this energy is
A chemical combustion.
B nuclear fusion.
C photosynthesis.
D solar radiation. (1 mark)

12 A compressed metal spring stores
A chemical potential energy.
B gravitational potential energy.
C kinetic energy.
D elastic potential energy. (1 mark)

13 An example of an extrusive igneous rock with a high silica content is
A granite.
B rhyolite.
C basalt.
D obsidian. (1 mark)

14 The metamorphic rock formed from sandstone is
A quartzite.
B gneiss.
C slate.
D marble. (1 mark)

(cont.)

15 An example of a trace fossil is
- **A** the tooth of a sabre-tooth tiger.
- **B** a mosquito in amber resin.
- **C** a dinosaur footprint.
- **D** petrified wood. (1 mark)

PART B *Restricted-response questions* 10 marks

Insert the missing word to complete each sentence.

16 The cell __________ allows various nutrient molecules to enter the cell and to allow waste materials to leave the cell. (1 mark)

17 Plants that are __________ by insects often have brightly coloured flowers with a strong scent to attract them. (1 mark)

18 The process of __________ is where the particles of matter spread out from a zone of high concentration to a zone of low concentration. (1 mark)

19 On the __________ the crust is about 35 km thick, but it is thinner under the oceans, roughly 5 to 10 km in thickness. (1 mark)

20 During ice cracking of rocks, as the water freezes, it __________ and exerts a great pressure in the rock. (1 mark)

21 Chemical potential energy is __________ in fuels such as petrol and diesel as well as in the foods we eat. (1 mark)

22 The __________ of chemical potential energy into heat energy is commonly observed in chemical reactions. (1 mark)

23 Igneous rocks form when __________ or lava cools and solidifies. (1 mark)

24 A hypothesis is a broad, general __________ of a collection of observations that predict a result and suggest a way of testing it. (1 mark)

25 Mild burns can be treated by holding the burnt area under __________ running water or by applying ice. (1 mark)

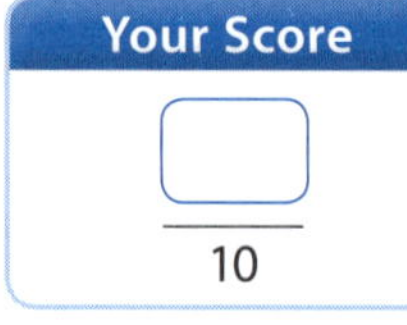

PART C *Written knowledge and skill questions* 25 marks

26 The following diagram shows hot magma making contact with some rocks leading to metamorphic changes to the parent rocks.

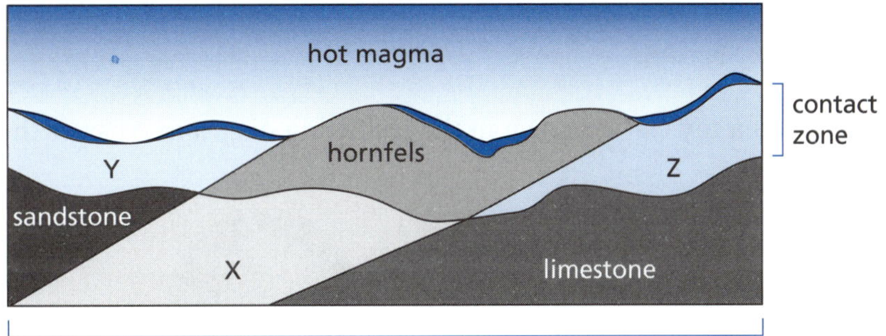

a Identify the type of metamorphic change illustrated. (1 mark)

b Name rocks X, Y and Z. (3 marks)

c Classify rock X. (1 mark)

d Compare the appearance of rock Y with the original sandstone. (1 mark)

27 The following text contains information about the extraction of copper minerals from the ground. Read the text and answer the questions that follow.

> Along with metals such as silver, lead and zinc, copper minerals are often located in hydrothermal deposits. Over a long period of geological time, high-pressure hot water associated with magma dissolved vast quantities of minerals to form concentrated solutions. These solutions seeped through cracks in the bedrock and, as they cooled, minerals crystallised and deposited to form a mineral vein or lode. One such mineral is chalcopyrite ($CuFeS_2$). When this mineral is exposed (through the erosion of overlying rock) to atmospheric air and water, it weathers to form other copper minerals such as malachite and azurite. It was these weathered minerals that were first mined by the ancient people of the Middle East. Once these surface minerals were mined and the copper extracted, miners were forced to dig deeper to expose the chalcopyrite. Copper can be extracted from chalcopyrite by crushing, grinding and froth flotation and then smelting it in a high-temperature furnace. Silica and air are added to the smelter and the copper is released in molten form as the iron and sulfur atoms in the mineral more readily combine to form iron oxides and sulfur dioxides than does copper. The silica combines with iron oxide to form a glassy slag.

a Use the information in the text to explain the meaning of the word *hydrothermal*. (1 mark)

b Copy and complete the following sentences.

i A mineral vein or lode is formed when … (1 mark)

ii Chemical weathering of chalcopyrite leads to … (1 mark)

(cont.)

c Match each word or term in the left column with the appropriate word or term in the right column. (2 marks)

A	azurite	i	using a high-temperature furnace
B	silica	ii	hot water leaching
C	smelting	iii	weathered mineral
D	magmatic heat	iv	slag

28 Students constructed a toy car as shown in the following diagram. Answer the questions that follow.

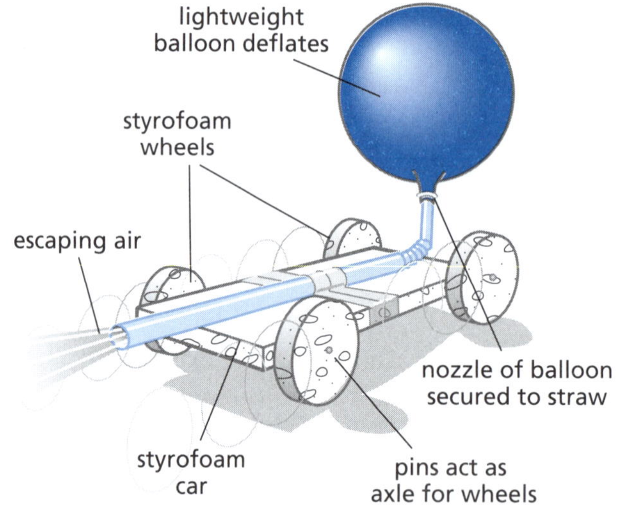

a Identify the type of energy stored by the inflated balloon. (1 mark)
b State the energy transformation that occurs as the balloon deflates. (1 mark)
c Suggest a reason why the toy car is made of styrofoam. (1 mark)

29 The following diagram shows a track on a roller-coaster. A carriage of mass 500 kg starts at point A on the track. Answer the following questions about its subsequent motion, assuming that there are no energy losses due to friction.

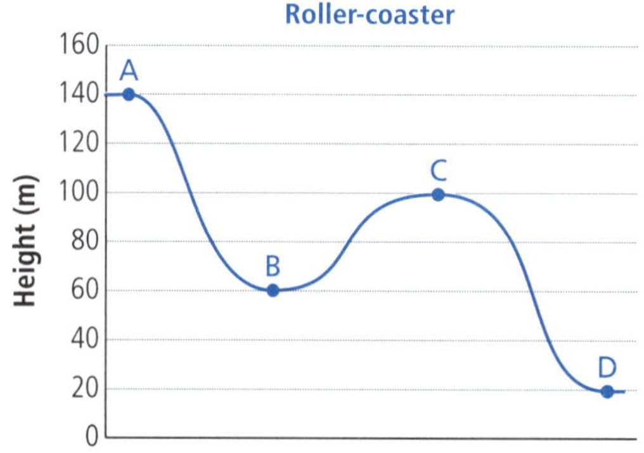

a Along which sections of the track will its velocity increase? (1 mark)
b Along which sections will its velocity decrease? (1 mark)

c Calculate the decrease in gravitational potential energy from point A to point B. (1 mark)

d What will be the kinetic energy of the carriage at B? (1 mark)

e What will be the kinetic energy of the carriage at C? (1 mark)

30 Jenny decided to do a first-hand investigation on the rising or leavening of bread dough. She discovered in her second-hand research that bread-making flour contains long molecules called starch, gliadin and glutenin. When water is added and the dough is kneaded, the glutenin and gliadin swell up and form an elastic material called gluten. The gluten traps the carbon dioxide gas formed by the fermentation of the starch by the yeast and so the dough leavens (rises) and expands. Jenny investigated the conditions that changed the rate of this leavening process.

a Jenny measured the rate of leavening by measuring how the circumference of balls of dough changed over time. This is shown in the following diagram. Will this measurement be the dependent or independent variable in this experiment? (1 mark)

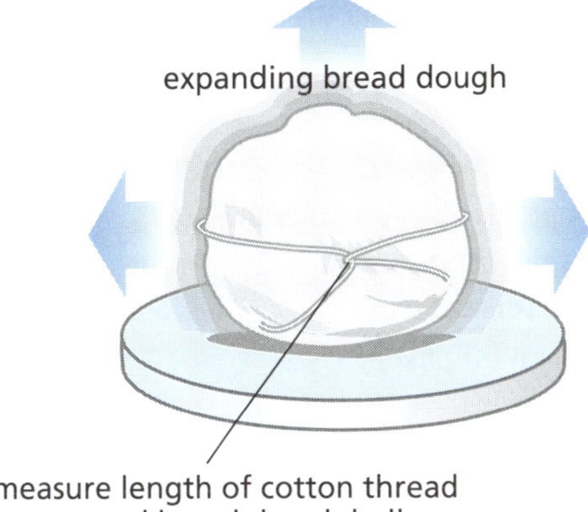

b Jenny investigated the effect of temperature on the leavening process. Describe how this investigation will be performed so that reliable results are obtained. (3 marks)

31 Sea snails extract minerals from the sea to make their shells. When they die, their skeletons collect at the bottom of the sea. Explain how these skeletons may ultimately form a bed of limestone.

(2 marks)

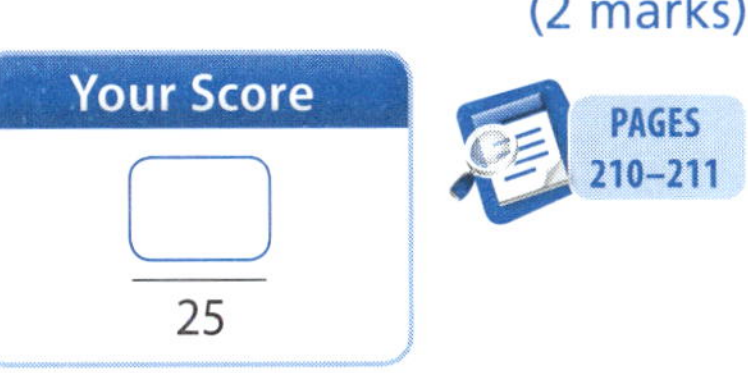

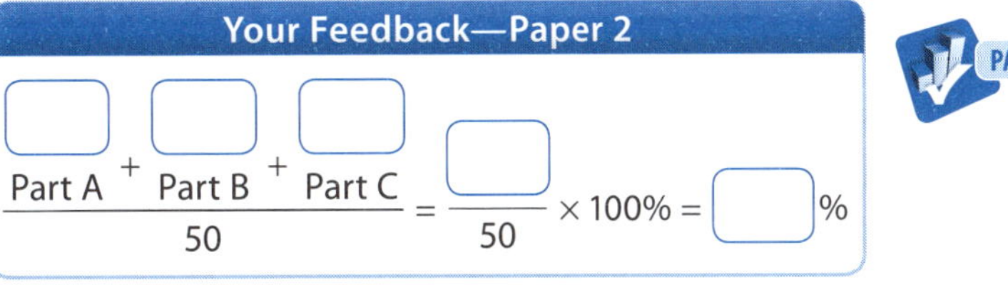

REVISION TESTS

Answers

CHECK YOUR ANSWERS

Strand: Biological sciences

MICROSCOPES
Cells Pages 4–5

1 **a** 10 × 10 = 100 times ✓
b **i** 3 mm ✓
ii 3 × 1000 = 3000 micrometres ✓
c size of seed = $\frac{3}{9}$ = about 0.333 mm = 333 micrometres ✓
d **i** 20 × 10 = 200 times ✓
ii 1.5 mm ✓ (as the high power field has twice the magnification of the low power field)

2 A monocular microscope has one eyepiece, while a binocular microscope has two eyepieces. ✓ Magnification with a monocular microscope is generally greater. ✓ In a monocular microscope you observe the specimen with transmitted light. With a binocular microscope you observe with reflected light. ✓

3 **a** crab larva ✓; **b** stylonychia ✓

4 A/2 ✓; B/3 ✓; C/5 ✓; D/1 ✓; E/4 ✓

5 4, 2 ✓, 1, 3 ✓

6 If you are focusing down you may damage the slide and specimen as the objective lens could crash down into the glass. This sort of accident cannot occur if you focus upwards and away from the specimen. ✓

CELL STRUCTURE
Cells Pages 10–11

1 **a** phospholipids, cholesterol, proteins and carbohydrates ✓
b They recognise foreign materials. ✓
c They allow passive diffusion of water-soluble molecules across the cell membrane. ✓
d Membranes allow cells to exchange substances with their external environment. ✓
e **i** cholesterol ✓; **ii** cellulose ✓

2 **a** xylem ✓; **b** nerve cell ✓; **c** guard cells ✓; **d** white blood cell ✓

3 **a** nucleus ✓; **b** mitochondrion ✓; **c** chloroplast ✓; **d** cell wall ✓

4 They cannot photosynthesise as they have no chloroplasts ✓, and their cell wall contains chitin instead of cellulose ✓.

5 **a** X = nucleus ✓; Y = cytoplasm ✓
b a unicellular organism belonging to the kingdom protista ✓
c The amoeba extends its cytoplasm outwards to surround the food. ✓ Once it is completely surrounded, the food is captured and then a food vacuole forms around it and digestion can begin. ✓

6 **a** stomate ✓
b **i** oxygen ✓, water vapour ✓
ii carbon dioxide ✓
c guard cells ✓
d Xylem cells form tubes through which water is transferred throughout the plant. In these veins, water evaporates into the air space and then out through the pore. ✓
Phloem cells form tubes that transport nutrient solutions throughout the plant. These nutrients are produced in the green leaf cells and then diffuse into the phloem cells to be transported. ✓

CELL DIVISION AND REPRODUCTION
Cells Pages 15–16

1 **a** As a cell gets larger, the ability of the cell membrane to transport nutrients into the cytoplasm and all of its organelles ✓, and to carry wastes out, cannot keep up with how fast the cell itself is growing ✓. (In other words, the surface-to-volume ratio of the cell decreases as it grows.) By dividing, and forming two smaller cells, this stress is relieved. ✓
b growth ✓, repair ✓ and asexual reproduction ✓

c Yes ✓ Mitosis occurs constantly, and in some cells more frequently than in others.

2 P = body (somatic) cells ✓; Q = sex cells (gametes) ✓; R = growth, repair, asexual reproduction ✓; S = sexual reproduction ✓; T = same as for parent ✓; U = half the number of the parent ✓; V = same as for parent ✓; W = half genetic material, with new gene combinations ✓

3 a 2 ✓ b 16 ✓

4 when they become too large ✓ There are some large cells. The ostrich egg is often cited as being a large cell (possibly meaning heavy; 15 cm long and weighing 1.4 kg). But so are some nerve cells especially in long animals, such as the giant squid and colossal squid, which may have nerve cells as long as 12 m.
In humans, the longest nerve cells are about 1.5 m, running from the base of the spine to the toes.

5 a chromosome ✓
b so that two new cells can be formed ✓

6 a false ✓ During meiosis the resulting gametes have half the number of chromosomes as the parent cell.
b true ✓ Two chromosomes may exchange fragments by a process called crossing over. This results in chromosomes that are mixtures of the original two chromosomes. Independent assortment is the random arrangement of pairs of chromosomes. These processes promote variation. Variation is necessary for species to become adapted to their environment and enables them to change when the environment alters.
c true ✓ Sex cells (gametes) are produced during meiosis.
d true ✓ Human cells normally contain 46 chromosomes (sperm cells 23 and egg cells 23). When sperm and egg combine they form a new individual: 23 + 23 = 46 chromosomes.
e false ✓ The two processes are not the same. For example, in mitosis the chromosome number is not halved.
f false ✓ The daughter cells contain half the number of chromosomes as compared to the parent cells in meiosis. Moreover, their chromosomes are different because of crossing over.
g true ✓ Each cell undergoing meiosis generates four daughter cells.

7 A = resting phase where the cell has its full number of chromosomes ✓
B = chromosome number doubles ✓
C = nuclear membrane disappears and microtubules attach to chromosomes ✓
D = chromosomes align in the middle of the cell ✓
E = the separated chromosomes are pulled apart ✓
F = microtubules disappear and the cell begins to divide ✓
G = two daughter cells produced, each with the same number of chromosomes as the parent cell ✓

DISEASE
Cells Pages 22–23

1 a There were many people living in close proximity so microbes could easily be spread from one person to another. ✓
b Today we know more about diseases ✓, what they are and how they spread, and we have drugs to deal with them ✓.

2 a bacterium ✓
b Direct: coughing/sneezing in close proximity to another individual. ✓
Indirect: through organism remaining in the air/dust or spread over large distances. ✓
c It has a protective coat to help withstand drying out and the effects of disinfectants. ✓

d *(any one)* many people living very closely to each other; poor hygiene; lack of medical facilities and medicines ✓

e Overcrowding: pathogen can be delivered directly ✓, not risking the chance it might be destroyed by it remaining outside the host for long periods. Malnutrition: weakened individuals are more susceptible to catching the disease ✓ as their immune system may not be working optimally.

f antibiotics ✓ (However, effective tuberculosis treatment is difficult, due to the unusual structure and chemical composition of the mycobacterial cell wall, which hinders the entry of drugs making many antibiotics ineffective.)

3 a Urine constantly flows down the urinary tract. Any pathogen would have to swim very effectively against this flow to reach the kidneys. ✓

b The majority of pathogens in food are killed by acid and enzymes as soon as they reach the stomach. ✓ Some survive the stomach environment to reach the large bowel. Here the pathogen has to compete with the many millions of faecal bacteria that normally live here. The chances of pathogens surviving this second barrier are small.

c Earwax is secreted in the ear canal of humans and other mammals. It protects the skin of the human ear canal, assists in cleaning and lubrication and also provides some protection from bacteria, fungi, insects and water ✓, partly because it is sticky. Its antimicrobial properties are due mainly to the presence of saturated fatty acids, and to the slight acidity of cerumen (pH typically around 6.1).

4 a The immune system has a sensory system that can detect when infection has taken place. It has a way of recognising self from non-self. Microbes and other particles are recognised as non-self. ✓ Pattern recognition molecules recognise infection and alert the innate immune system, which then uses a range of systems to attack the infection. It is for this reason that people who have had organ transplants, such as heart, kidneys or liver, have to take anti-rejection drugs to prevent their bodies from rejecting these foreign parts.

b White blood cells are continually on the lookout for signs of disease. When a germ does appear, these cells have a variety of ways by which they can attack. Some will produce protective antibodies that will overpower the germ. ✓ Others will surround and devour the bacteria ✓ and other debris.

5 No, antibiotics do not work against these viruses. ✓

6 Infectious diseases are caused by pathogens ✓, which are disease-causing organisms. The major groups of pathogens are viruses, bacteria, fungi, protozoans and multicellular parasites. They can be transmitted from one person to another ✓. Non-infectious diseases are not caused by pathogens ✓ and cannot be spread from person to person ✓.

7 a environmental ✓

b nutritional deficiencies ✓

c nutritional deficiencies ✓

d genetic disorders ✓

e environmental ✓

f environmental ✓

g genetic disorders ✓

8 a false ✓ While colds and flu have similar symptoms, the flu (influenza) makes you feel much worse. Both colds and flu are caused by viruses.

b false ✓ Antibiotics are an effective treatment for bacterial infections, but are not effective against viruses. There are antiviral medications that can be taken for influenza, but none for the common cold.

- **c** false ✓ For antibiotics to be effective, you must take the correct dose for the length of time prescribed. Even though your symptoms have improved, the bacteria that caused the illness may not yet have been completely destroyed.
- **d** true ✓ However, meningococcal bacteria are not able to live for very long outside the body, so the disease is not easily spread from one person to another. It is not as contagious as colds or flu, and is not usually spread by casual contact or by breathing the air.
- **e** false ✓ This was commonly believed to be the reason but it is now known that during cold weather people are congregated indoors more often and so the common cold can more easily spread because of this close proximity.
- **f** true ✓ Washing hands regularly, and especially after going to the toilet, prevents germs from spreading.
- **g** false ✓ The bugs are called pathogens. Parasites are organisms that live in or on another organism (its host) and benefit by deriving nutrients at the host's expense.
- **h** true ✓ Mesothelioma is a rare, and often fatal, cancer of the mesothelium, the membrane covering most of the body's internal organs. The disease can develop decades after exposure to asbestos.

9
- **a** 1956 ✓
- **b** In 1968 ✓ around 8500 cases ✓ were reported.
- **c** soon after 1996 ✓
- **d** around 1992 ✓
- **e** less than 1000, say, about 500 ✓
- **f** There has been a dramatic decrease in the incidence of measles. ✓

MULTICELLULAR ORGANISMS

Body systems Pages 27–28

1
- **a** **i** A ✓; **ii** phloem (B) ✓
- **b** C ✓ cambium ✓
- **c** 22 years ✓ It is in its 23rd growing year.
- **d** B (and A) ✓ (The cork layer, A, is dead but the living phloem tissue, B, must be cut to ringbark the tree.)
- **e** xylem ✓ Xylem transports water from the root via the stem to the leaves so that all cells are hydrated. ✓

2
- **a** A tissue is made up of cells that have a similar structure and function. ✓
- **b** *(Other answers are possible.)* Muscle cells ✓ make up tissue called muscle ✓.

3 The aim of the experiment is to show which tissues in the celery stem conduct water. ✓ The experiment shows that the xylem tissue is confined to sections of the stem (called vascular bundles) rather than the whole of the stem. ✓

4
- **a** false ✓; **b** true ✓; **c** false ✓; **d** true ✓;
- **e** true ✓; **f** true ✓; **g** true ✓; **h** true ✓;
- **i** true ✓; **j** false ✓

5
- **a** See the graph below.

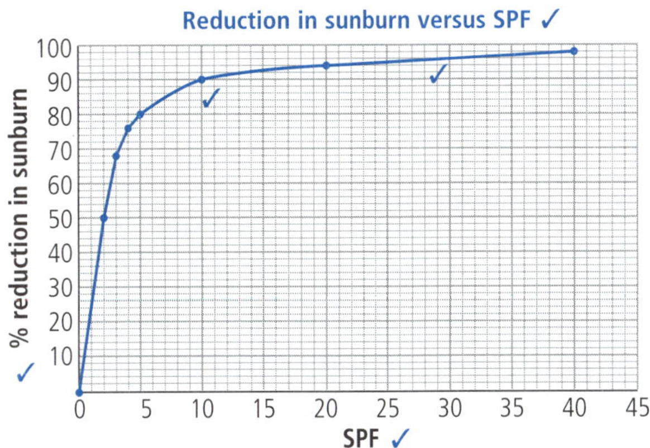

- **b** about 93% ✓
- **c** *(Other answers are possible.)* Keep out of the Sun, especially between 11 am and 2 pm. ✓ Wear a hat with a broad brim and a shirt with long sleeves. ✓

REVISION TESTS
Answers

CHECK YOUR ANSWERS

DIGESTIVE SYSTEM
Body systems Pages 32–33

1 **a** X = incisors ✓; Y = molars ✓
b herbivore ✓
c sheep ✓

2 Digested food nutrients diffuse into the blood vessels from the small intestine ✓ and then the nutrients diffuse into the cells ✓.

3 **a** large intestine ✓; **b** duodenum ✓;
c stomach ✓; **d** mouth ✓;
e small intestine ✓

4 A = mouth ✓; B = oesophagus ✓;
C = stomach ✓; D = pancreas ✓;
E = large intestine ✓; F = small intestine ✓;
G = gall bladder ✓; H = liver ✓

5 **a** This occurs in the mouth. ✓ Chewing and mixing with saliva begins the process ✓ by lubricating and binding chewed food, and wetting and partly dissolving the food ✓. Enzymes in saliva begin starch digestion. ✓

b **i** This allows a rather large meal to be consumed quickly and dealt with over a period of time. ✓

ii While starch digestion was begun in the mouth the really big job, particularly the digestion of proteins, begins in the stomach. ✓

iii This mixes and grinds foodstuffs with gastric juices, resulting in liquefaction of foods, a necessary preliminary step before delivery to the small intestine. ✓

iv As the food is liquefied in the stomach, it is slowly released into the small intestine for further processing and this allows for maximum absorption of the nutrients. ✓

6 *(Other answers are possible.)* The hydra has no anus as a separate opening to the mouth ✓; there are no organ systems feeding juices into the digestive cavity ✓.

CIRCULATORY AND RESPIRATORY SYSTEMS
Body systems Pages 37–38

1 X = vein ✓; Y = capillary ✓; Z = artery ✓

2 **a** pulmonary artery ✓; **b** aorta ✓;
c vena cava ✓; **d** pulmonary vein ✓

3 **a** to allow the oxygen/carbon dioxide to diffuse in and out of the cell membranes with which they communicate ✓

b to protect the body against pathogenic microbes and disease ✓

4 **a** See the table below. ✓

Chamber/vessel	Blood pressure range (mm of mercury) high–low
right ventricle	28–0
pulmonary artery	28–10
pulmonary vein	2–minus 2
left ventricle	120–0
main arteries of the body	120–80
capillaries in body tissues	35–8
veins	8–minus 2

b See the graph below.

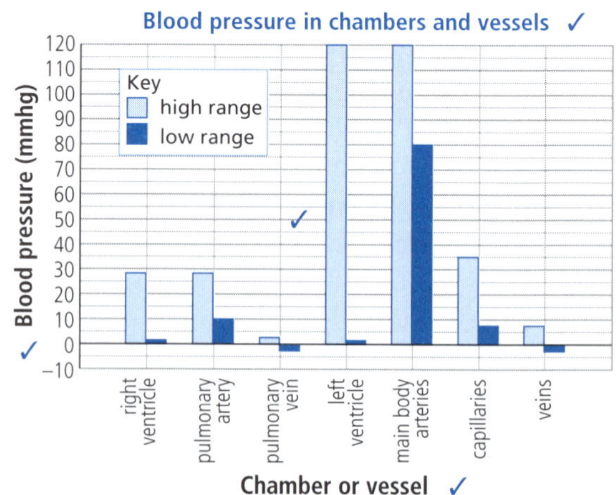

c main arteries ✓
d pulmonary vein/vein ✓
e Yes, it exerts less force on the blood so blood going to and from the lungs is at a lower pressure than that of the body. ✓

- **f** The blood pressure in the veins is low so the blood is moved back to the heart via nearby muscle contractions. ✓
- **g** the arteries ✓
- **h** The capillaries would rupture (burst) as they are thin. ✓

5 **a** false ✓; **b** true ✓; **c** false ✓; **d** false ✓; **e** false ✓

6 **a** trachea (windpipe) ✓; **b** alveoli ✓; **c** diaphragm ✓; **d** breathing in ✓

EXCRETORY AND MUSCULOSKELETAL SYSTEMS
Body systems Pages 43–44

1 *(any three)* Kidney activity is controlled by a person's choices, the environment as well as hormones. For example, if a person consumes large amounts of protein, much urea will be in the blood from the digestion of the protein. This will need to be eliminated. Also, on a hot day a body will retain water for sweating and cooling, so the amount of urine is reduced. On the other hand, drinking lots of water and other drinks will produce much urine. Certain medication, such as diuretics, can increase the excretion of water from bodies. ✓✓✓

2 As you grow older, small bones join together to make big ones. ✓

3 **a** **i** $5 \times 12 = 60$ L ✓
ii $60 \times 24 = 1440$ L ✓
b $\frac{1.5}{1440} \times 100 = 0.1\%$ ✓

4
- **a** V ✓; femur ✓
- **b** U ✓; patella ✓
- **c** Q ✓; clavicle ✓
- **d** P ✓; cranium ✓
- **e** T ✓; vertebrae ✓
- **f** R ✓; scapula ✓
- **g** S ✓; sternum ✓

5 **a** The muscles which we can control consciously are called the voluntary muscles ✓ and the ones we can't control are the involuntary muscles ✓.
- **b** voluntary: swivelling the head ✓, raising your leg ✓; involuntary: heart ✓, the movement of your diaphragm (for breathing) ✓, blinking your eyes, stomach muscles churning food
- **c** through signals from your nerves ✓

6 **a** upper arm: humerus ✓; lower arm: radius ✓ and ulna ✓
- **b** hinge joint ✓
- **c** They are connected to bones by tough, cord-like tissues called tendons ✓, which allow the muscles to pull on bones.
- **d** Raise your forearm by relaxing the triceps ✓ and contracting the biceps ✓. Lower your forearm by relaxing the biceps ✓ and contracting the triceps ✓.

7 **a** the urinary system ✓
- **b** W = kidney ✓; X = ureter ✓; Y = urinary bladder ✓; Z = urethra ✓
- **c** One of the major functions of the urinary system is the process of excretion. ✓ Excretion is the process of eliminating waste products of metabolism ✓ and other materials that are of no use to an organism. The urinary system maintains an appropriate fluid volume by regulating the amount of water that is excreted in the urine. ✓

8 **a** $H = 1.880 \times 45.5 + 81.31 = 167$ cm ✓
b $H = 1.945 \times 45.5 + 72.84 = 161$ cm ✓

ASEXUAL AND SEXUAL REPRODUCTION
Body systems Pages 48–49

1 **a** asexual reproduction ✓, as only one parent is involved ✓
b A The amoeba stops moving and becomes more rounded. ✓
B The nucleus begins to divide. ✓

C After the nucleus has finished dividing, the cytoplasm begins to constrict. ✓

D The constriction continues to divide the cytoplasm. ✓

E The two halves, each with a nucleus, move slightly apart and have almost separated. ✓

F The daughter amoebae have completely separated. ✓

2 A Make a slanting incision into the stem with a sharp knife, and keep the cut open using a matchstick or toothpick. ✓

B Wrap and tie with moist sphagnum moss. ✓

C Wrap the whole moss with a plastic film or waterproof bag to prevent water loss from the moss. ✓

D After some time, roots should start to appear through the moss. ✓

E When a good root system has formed, sever the rooted tip from the stock plant and pot it. ✓

3 Asexual reproduction can be advantageous to certain plants and animals, especially those that remain in one particular place and are unable to look for mates. ✓ Also numerous offspring can be produced without requiring a great amount of energy or time. ✓ (Think of how quickly a few bacteria can multiply once inside you causing disease.) Environments that are stable with very little change are the best places for organisms that reproduce asexually ✓, as these organisms are identical to the parent and don't exhibit variation. This lack of genetic variation can be a disadvantage as all of the organisms are genetically identical and therefore share the same weaknesses. ✓ If the stable environment changes, or some disease is introduced, the consequences could be deadly to all of the individuals. ✓ As some offspring may not be able to move far, they can be close together and compete for food and space. ✓ (Consider potato plants that grow close together and struggle for soil, nutrients and light. This will consequently cause the plants to be less healthy.) Overall, asexual is faster and easier than sexual reproduction, because another partner is not needed, and doesn't have to be found.

4 **a** Self-pollination is the transfer of pollen from an anther to a stigma of the same flower, or to another flower on the same plant. ✓ Cross-pollination occurs when pollen is delivered to a flower from a different plant. ✓

b Both gametes come from the same parent so there is little variation. ✓ Inbreeding can result and if there are unfavourable characteristics in the parent, they are much more likely to be exposed than if the plant cross-pollinates. ✓

c Much pollen needs to be produced of which only a small fraction will ever reach another plant. ✓ In self-pollinating plants the pollen doesn't need to travel that far. Cross-pollination also assumes there is another plant of the same species nearby ready to be pollinated. ✓ (But it is genetically favourable because genes are transferred and variation increases.)

d Pollen can be spread by the wind or animals. These are usually insects (generally bees, butterflies, moths, flies and beetles) that transport the pollen of plants when using the flowers for feeding, breeding or hiding. When attracted by these mechanisms ✓, the insect rubs or touches the stamens (male reproductive part) and is covered in pollen. Then the insect moves to another flower and the pollen on the insect reaches the stigma (female reproductive part) and pollinates it. ✓ Honey bees are very important insect pollinators. Mostly, both bees and the plants they visit benefit. The honey bee gets some food and the plant gets pollinated.

e Although the plant and the insect may benefit because of their relationship with each other, the insect visiting a flower usually does not intentionally pollinate the flower. ✓

5

Letter	Name	Function
V ✓	ovary ✓	contains egg cells
P ✓	petal ✓	attracts insects and other pollinators
Q ✓	stigma ✓	traps pollen
R ✓	anther ✓	makes the pollen
S ✓	sepal ✓	used to protect the growing flower bud
U ✓	style ✓	pollen travels through this part of the pistil
T ✓	filament ✓	part of stamen providing support

6 Male

Letter	Name	Function
E ✓	penis	the organ used to deposit semen into the vagina of the female during intercourse
F ✓	urethra	a tube running through the length of the penis that carries semen; can also discharge urine from the bladder
D ✓	scrotum	the external sac holding the testes
C ✓	testes	site of sperm production; located within the scrotum
B ✓	vas deferens	also called the sperm duct, it is a tube carrying sperm away from the testes
A ✓	seminal vesicles	one of several glands producing protective and nutrient fluids for the sperm; feeds into the sperm duct

Female

Letter	Name	Function
I ✓	ovaries	site of egg production, located one on each side
G ✓	vagina	canal into which semen is deposited and the birth canal through which the baby is born
J ✓	fallopian tubes	also known as oviducts, the tubes through which mature eggs pass following ovulation
K ✓	uterus	also known as the womb, this is a muscular organ where the fertilised egg will implant and grow to form the baby
H ✓	cervix	a narrow entryway between the vagina and uterus; muscles here are flexible expanding to allow baby to pass during birth

BIOTECHNOLOGY
Body systems Pages 53–54

1 **a** tissue culture or micro-propagation ✓

b P Tissue samples are cut or scraped from the parent plant. ✓

Q Tissue samples are placed in a nutrient growth medium. ✓

R The samples grow into small plants. ✓

S When suitably large, with a good root system, these plantlets are planted in a good potting mixture. ✓

2 **a** The cloning of animals has many important commercial implications. It allows an individual animal with desirable features, such as a cow that produces a lot of milk or a horse that has won many races, to be duplicated several times. ✓ Prize-winning animals can be cloned and those clones sent around the world.

b **i** D ✓; **ii** B ✓; **iii** F ✓; **iv** C ✓; **v** E ✓; **vi** G ✓; **vii** A ✓

3 *(any three)* produce multiple copies of identical plants in a short period; produce plants with desired features; produce plants any time regardless of climate; produce plants without seeds; produce healthy plants from plants that are partly infected with disease ✓✓✓

4 *(any three)* artificial insemination; in-vitro fertilisation; somatic cell nuclear transfer; reproductive cloning; cryopreservation; fertility medications ✓✓✓

5 a To ovulate means to produce an egg and so the lizards are female. ✓ However, these lizards can also reproduce by sexual means when there is a male lizard about.
  b Hybrid species are more prone to extinction because they don't produce much genetic diversity (variation) from generation to generation. ✓ Genetic diversity keeps a species viable and healthy in the long term, especially if there are environmental changes. ✓

6 a in vitro fertilisation (IVF) ✓(the process of fertilising eggs with sperm outside of the human body)
  b Hormones are usually given to stimulate the ovaries to produce more than the usual one egg per cycle so several can be collected at once. When it is the appropriate time to collect the eggs, a fine needle is passed through the vaginal wall and into the ovaries. ✓ A few hours after collection, the man provides a semen sample. ✓ The eggs are mixed with the sperm. ✓ If an egg is fertilised by a sperm, a zygote develops. At an appropriate size, one or two zygotes are transferred back to the woman's uterus ✓ at the right time in her menstrual cycle. Only one or two embryos are transferred back at any one time.

7 a Immunosuppressants are powerful medicines that dampen down the activity of the body's immune system ✓. As a transplanted kidney is recognised as non-self, the body's immune system kicks in to reject it as it would do with any foreign organic matter. So patients need to take these drugs to prevent kidney rejection. ✓
  b bacteria ✓
  c As the patient is taking drugs to prevent rejection, the body's immune system is suppressed so it is easier for germs to infect the patient. ✓

Strand: Chemical sciences

MATTER AND CHANGES OF STATE
Elements and compounds Pages 59–60

1 a sand = 260.00 g ✓ water = 100.00 g ✓ oxygen = 0.15 g ✓
  b Sand has the greatest density. ✓ Oxygen has the least density. ✓

2 liquid ✓

3 a The plunger will not move ✓ as the water is incompressible ✓.
  b The plunger will move in readily ✓ as gases are readily compressed due to the large spaces between particles ✓.

4 The Martian air has fewer gas particles than air on Earth. ✓

5 a diffusion ✓
  b The chlorine particles and the air particles are in constant motion. They collide with each other as they move about and eventually these collisions allow the chlorine particles to spread throughout the vessel. ✓

6 $D = \frac{m}{V}$ ✓ $= \frac{52.5}{5.0} = 10.5\ \text{g/cm}^3$ ✓

7 a liquid ✓
  b i The mercury vapour would turn from a gas to a liquid. ✓
    ii condensation ✓
    iii solidification ✓

REVISION TESTS

Answers

CHECK YOUR ANSWERS

ELEMENTS

Elements and compounds Pages 64–65

1 **a** to show whether electricity is flowing and therefore which elements conduct ✓

b any metal (e.g. copper, nickel) ✓

c zinc ✓, iron ✓, nickel ✓, calcium ✓

2 **a** Ni ✓, Pb ✓

b mass = 62 + 208 – 1 = 269 u ✓

c p = 28 + 82 = 110 ✓

d Z = p = 110 ✓

3 A/4 ✓; B/3 ✓; C/2 ✓; D/7 ✓; E/8 ✓; F/5 ✓; G/1 ✓; H/6 ✓

4 A = density greater than 2 ✓; B = density less than 2 ✓; C = Co, Fe ✓; D = W, Zn ✓; E = sodium ✓; F = magnesium ✓; G radioactive form used in cancer treatment ✓; H = used in industry ✓; I = Co ✓; J = Fe ✓; K = tungsten ✓; L = zinc ✓

COMPOUNDS

Elements and compounds Pages 70–71

1 **a** pure ✓; **b** ion ✓; **c** compound ✓; **d** formula ✓; **e** polymer ✓

2 **a** true ✓ Compounds have a fixed number of elements of each type. For example, water always has two atoms of hydrogen and one atom of oxygen

b false ✓ The properties of the compound can now be vastly different from the individual elements which constitute it. Elements in a compound do not retain their individual properties.

c false ✓ They can be made from the same element; for example, chlorine gas (Cl_2) or ozone (O_3).

d true ✓ Separating aluminium or titanium, for instance, from compounds in their ores takes a tremendous amount of energy.

3 In a compound the proportions of each element present are constant (always the same). ✓ For example, in water (H_2O) a molecule always has two atoms of hydrogen combined to one atom of oxygen. ✓ It doesn't matter whether the sample of water came from a chemical reaction or from the deepest parts of the ocean.

4 **a** lead ✓, sulfur ✓ and oxygen ✓

b 1:1:4 ✓

5 **a** compound ✓

b carbon ✓, hydrogen ✓ and oxygen ✓

c 45 ✓ (= 12 + 22 + 11)

d No ✓ Even though it contains the same elements, they are not in the same proportion (ratio) ✓ as for sucrose. (This compound is another type of sugar, called glucose.)

e C:H:O = 6:12:6 = 1:2:1 ✓

6 **a** carbon, hydrogen and oxygen ✓

b $p = 3$ ✓, $q = 6$ ✓, $r = 1$ ✓

c chemical bonds ✓

d covalent bonds ✓ as these join together non-metal and non-metal elements ✓

7 **a** Dyes are added so one fuel can easily be distinguished from another fuel ✓ without complicated chemical analysis.

b The dyes used have to be soluble in the fuels ✓ and they must not interfere with the proper functioning of the fuel ✓.

8 **a** A chemical bond is an attraction between atoms that allows the formation of chemical substances containing two or more atoms. ✓

b A = two atoms share their electrons ✓; B = one atom transfers its electron to another atom ✓; C = a molecule forms ✓; D = ion ✓; E = negative ion ✓; F = covalent ✓; G = ionic ✓

9 **a** carbon = 4 ✓; hydrogen = 10 ✓; oxygen = 1 ✓

b carbon = 4 ✓; hydrogen = 10 ✓; oxygen = 1 ✓

c The arrangement of the elements in the two compounds are different. ✓

10 a polymerisation ✓; b plastics ✓

SIMPLE CHEMICAL REACTIONS
Elements and compounds Pages 75–76

1 a a physical change ✓; no new substance is formed ✓
b While this is true of many chemical reactions, compared to physical changes, it does not define a chemical change. In a chemical change, a new substance is formed. ✓ So while Humpty can't be reconstituted, the same bits and pieces are all there as they were before his fall ✓ only in a slightly disarranged form.

2 Matter is never destroyed or created in chemical reactions. ✓ The same number of particles that existed before the reaction exist after the reaction.

3 a Q ✓ The candle is melting and dribbling down the side.
b reduction in the size of the candle ✓ (indicating it is being consumed), heat generated ✓, light produced ✓

4 a iron ✓; b magnesium ✓; c bromide ✓; d sodium ✓ iodide ✓

5 nitrogen + hydrogen → ammonia ✓✓

6 a carbonic acid → carbon dioxide + water ✓✓
b The bubbles of carbon dioxide have had time from the slowly decomposing carbonic acid to work their way out of the solution. ✓

7 a oxygen ✓
b chloride ✓
c iron ✓, sulfide ✓

8 a sodium oxide ✓, carbon dioxide ✓
b potassium chloride ✓, oxygen ✓

9 The same ✓ as the Law of Conservation of Matter says matter cannot be created or destroyed ✓.

10 a silver chloride → chlorine + silver ✓
b The silver which forms deposits as fine metal. ✓ This gives the grey-to-black appearance.
c The chlorine produced in the reaction escapes into the atmosphere as it is a gas. ✓

Strand: Earth and space sciences

EARTH STRUCTURE AND MINERALS
Minerals and rocks Pages 80–81

1 a fingernail ✓
b steel file ✓
c fingernail, calcite, glass ✓
d copper coin, calcite, pocket knife blade, steel file, quartz, topaz ✓
e i *haemo* = blood (reddish) ✓
ii The streak shows true colours when all the mineral is ground to the same size. Otherwise its colour is variable. ✓

2 a $D = \frac{m}{V} = \frac{27}{10} = 2.7 \text{ g/cm}^3$ ✓
b $D \text{ (mineral)} = \frac{m}{V} = \frac{33.6}{8.0} = 4.2 \text{ g/cm}^3$
Therefore, the mineral is chalcopyrite (fool's gold). ✓

3 a hardness ✓
b galena ✓
c quartz ✓

4 a crust ✓
b lithosphere ✓
c oxygen and silicon ✓
d The mineral was not a carbonate. ✓
e 3 ✓
f false ✓
g false ✓

5 a 100 – (46.6 + 27.7 + 8.2 + 5.0 + 3.6) = 8.9% ✓

b See the graph below.

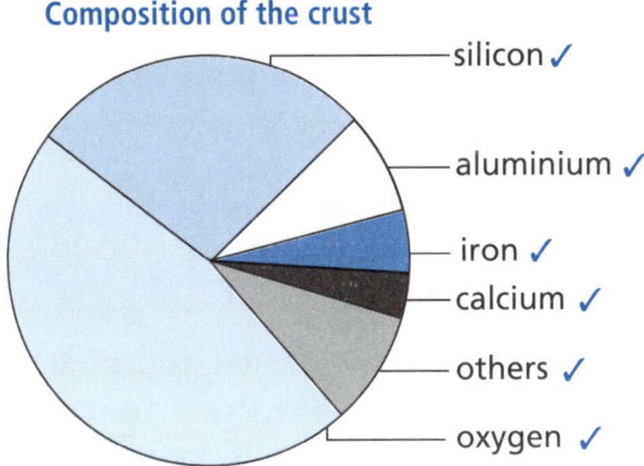

WEATHERING AND EROSION
Minerals and rocks Pages 86–87

1 gravity ✓

2 No, they will not remain in balance. ✓ Erosion will become greater than deposition. Waves lose most of their energy as they roll up the coast, but a sea wall deflects the wave (and its energy). As the wave bounces back, it erodes the beach. As the beach erodes even further, more force is unleashed against the sea wall. No matter how strong the sea wall is, it is only a matter of time before it, too, is eroded away by the sea. ✓

3 a true ✓; b true ✓; c true ✓; d false ✓; e false ✓

4 a wind ✓
b Yes ✓ Wind can pick up beach sand and this windblown sand can abrade surfaces. ✓

5 materials dissolved: b/mineral salts ✓
materials suspended: c/clay ✓, e/fine sand ✓
materials rolled/pushed along river bed: a/pebbles ✓, d/boulders ✓, f/very coarse sand ✓

6 a *(Many answers are possible.)* Wear safety glasses so that no liquids or particles can get into the eye ✓; do not touch the hot rock or the ends of the tongs ✓; steam is released when the rock is placed in the water so do not get scalded by the steam ✓.
b physical weathering due to temperature changes ✓
c Small pieces of sandstone from the rock surface may break off into the dish of water. ✓

7 a Soda water contains carbonic acid which is an agent of chemical weathering in nature. ✓
b A change in mass of the rock could be used as evidence of chemical weathering. Also the rock may change its appearance. ✓
c Five or more experiments should be performed in separate containers rather than just one experiment. ✓

8 The sinkhole will initially form as a depression in the limestone as the top of the island is attacked by carbonic acid which forms from carbon dioxide dissolving in water. The sinkhole will eventually form as the acid continues its attack deeper. ✓ The sea cave will form as waves weather and erode the limestone at sea level. ✓ Eventually these two eroded formations meet as shown in the diagram.

SEDIMENTARY ROCKS AND FOSSILS
Minerals and rocks Pages 91–92

1 a coal/limestone ✓ *(either one)*;
b cementation ✓; c plant ✓;
d conglomerate ✓; e fossils ✓; f clastic ✓;
g clay ✓; h calcium ✓; i weathering ✓;
j deposited ✓

2 P Fast-flowing streams erode rocks and carry particles of different sizes downstream. ✓
Q Where the river slows down, larger particles settle to the bottom. ✓
R When the river reaches the lake or ocean, sediments fall to the bottom under gravity. ✓

S The weight of overlying sediment compresses and compacts the rock. ✓

3 Sedimentary rock is often formed in layers because the original sediment is laid down in layers. ✓

4 **a** one of the many different types of particles that make up this rock ✓
b 4 ✓ (only count the regular shapes)
c the cementing material ✓
d Compaction forces the water out of the sediment. Salts in the water crystallise under pressure and cement the sediment particles together ✓ (like glue).

5 A Millions of years ago an animal died and its remains found their way to the bottom of a lake or sea where the soft parts were eaten or rotted away. ✓ The hard parts such as the skeleton, including the skull, remained behind.
B Layers of sediment are deposited and the skeleton is crushed by the weight of the overlying material. ✓
C While the sediment hardens into rock, parts of the bone dissolve away to be replaced by minerals seeping from the surrounding sediment. ✓
D Earth movements over a long time period raise the rock above sea level. ✓
E Erosion exposes the fossil, where it is found. ✓

6 **a** Soft internal organs, muscles and skin rapidly decay ✓ and are rarely preserved, but the bones and shells of animals stand a far better chance of fossilisation ✓. Almost no fossil record exists for soft organisms such as jellyfish and worms.
b There are so many shelled invertebrates, compared to larger land animals, that some stand a better chance of fossilisation. ✓ In order to become fossilised, land animals must die in a watery environment and become buried in the mud and silt. Because of this requirement, most land creatures never get the chance to become fossilised unless they die next to a lake or river. ✓

IGNEOUS AND METAMORPHIC ROCKS
Minerals and rocks Pages 96–97

1 **a** sedimentary ✓; **b** metamorphic ✓; **c** igneous ✓; **d** extrusive ✓; **e** intrusive ✓; **f** foliation ✓; **g** pressure ✓; **h** magma ✓; **i** Regional ✓; **j** sandstone ✓; **k** basalt ✓; **l** obsidian ✓; **m** pumice ✓; **n** metamorphic ✓; **o** melting ✓; **p** weathering ✓

2 **a** erosion ✓ Over a period of time the original surface has been worn down and transported away.
b contact metamorphic rock ✓ As the molten magma pushed its way into the country rock, the rock in close contact with it was altered by the tremendous heat. ✓✓
c It is harder ✓, as it stands above the surrounding country rock that eroded faster ✓. Metamorphic rocks are harder than other types of rock, so they are more resistant to weathering and erosion.

3 See the table below.

	felsic	intermediate	mafic
extrusive	rhyolite ✓	andesite ✓	basalt ✓
intrusive	granite ✓	diorite ✓	gabbro ✓

4 It needs to be strong enough to withstand the use as blades or as arrow heads. ✓ It also needs to be able to form sharp cutting edges. ✓

5 **a** The rock cycle is a natural process whereby rocks transform from one rock type into another rock type over time, a type of natural recycling. ✓ It is a model describing the formation, breakdown and reformation of a rock as a result of

sedimentary, igneous and metamorphic processes.

b A = igneous rocks ✓; B = sedimentary rocks ✓; C = metamorphic rocks ✓

c No, the cycle is more random. ✓ A sedimentary rock could be eroded away and reformed into another sedimentary rock. Alternatively, it could be buried and altered by heat and/or pressure giving a metamorphic rock, or it could be melted to then form an igneous rock. What actually happens depends on the circumstances at the time and what is occurring where the rock is found. ✓

6 X Intrusive igneous rocks are formed from magma that cools and solidifies within the crust of the Earth. Buried and prevented from cooling, the magma cools slowly ✓ and as a result these rocks are coarse grained ✓. The mineral grains in such rocks can generally be identified with the naked eye.

Y These igneous rocks are formed at a depth in between the plutonic and volcanic rocks. ✓ The crystal size is intermediate between coarse-grained and fine-grained igneous rocks. ✓ These form due to the cooling and resultant solidification of rising magma just beneath the Earth's surface. They often form in dykes, sills and laccoliths.

Z Extrusive igneous rocks cool and solidify far more quickly than intrusive igneous rocks. ✓ Since the rocks cool very quickly, they are fine grained. ✓ If the cooling has been so rapid as to prevent the formation of even small crystals after extrusion, the resulting rock may be mostly glass (such as obsidian).

7 **a** limestone ✓

b *(any two)* Marble is used for stone home furnishings, statues, buildings, tombstones, fireplace mantels, floor tiles, countertops, clocks, hotplates, tables, pillars, structural resurfacing, even bathroom appliances. ✓✓ Marble is soft and is easy to carve or cut into shapes. In Ancient Greece marble was used to make statues and buildings like the Parthenon. Sculptors and architects have used this stone for centuries.

c No. Whatever is mixed in with the limestone initially decides the final colour of the marble. ✓ If limestone is pure, then a white marble results. ✓ If it has haematite or clay in it, then a reddish-coloured marble forms.

d Resistance to abrasion ✓, as well as the hardness of the component minerals ✓, is important for floor and stair treads. It also needs to look good.

e Atmospheric moisture containing carbon dioxide and sulfur dioxide form acids that can eat away at the marble. More importantly, marble is especially susceptible to stains and etching from a variety of common liquids, such as fruit juice and coffee. Food acids, during the preparation of food, can slowly react with the marble top ✓ so polishing protects it from becoming pitted. A protective polish coat places a barrier between the food acids and the calcite. ✓

METALS AND ORES
Ores and environmental issues Pages 102–103

1 **a** A native metal is one that can be found uncombined in nature. ✓

b A quartz vein is formed when quartz minerals crystallise from hydrothermal solutions that penetrate through cracks in surrounding rocks, forming sheet-like structures. ✓

c *(any three)* malleable; lustrous; readily melted and cast; unreactive and does not corrode ✓✓✓

2 **a** Pyrite is bronze or yellow in colour whereas galena is black-grey in colour. ✓ (Alternatively, pyrite is much harder than galena.)

b Cassiterite crystals are tetragonal whereas chromite crystals are octahedral. ✓ (Alternatively, cassiterite is slightly harder than chromite.)

c Chalcopyrite forms brassy yellow crystals whereas molybdenite forms grey crystals. ✓ (Alternatively, chalcopyrite is much harder than molybdenite.)

3 **a** aluminium ✓

b iron ✓

c uranium ✓

4 **a** Hydrothermal solutions of gold and quartz form when ground water comes in contact with cooling magma or hot rocks. ✓ The hydrothermal solutions move towards the surface and crystallise in fissures and cracks in the rock. The solid gold is embedded in the quartz. ✓

b Erosion of the landscape exposes gold-bearing quartz. ✓ Rivers can transport gold fragments and where the water currents slow down, the gold is deposited to produce placer deposits of alluvial gold. ✓

5 **a** sedimentary deposit ✓

b Suspensions of insoluble materials gradually settle under gravity in calm bodies of water to form sediments. ✓ In the case of iron ore the iron ions have formed insoluble iron oxides when they react with oxygen. The layers of iron oxides usually alternate with other sedimentary deposits such as clay which is converted to shale by heat, pressure and cementation. ✓

c iron ✓

EXTRACTING METALS FROM COMMON ORES

Ores and environmental issues Pages 108–109

1 **a** metal ✓

b bauxite ✓

c aluminium oxide (or alumina) ✓

d The large amounts of electricity are used to heat the minerals until they melt and also to electrolyse the mixture to produce aluminium. ✓

2 **a** metal ✓

b chalcopyrite ✓, chalcocite ✓ (or bornite)

c froth flotation ✓

d *(any one)* copper electrical wires, bases of saucepans, used to produce brass or bronze ✓

3 **a** aluminium ✓

b alumina ✓, cryolite ✓

c to ensure that the mixture of alumina and cryolite are fully melted ✓; to ensure the aluminium that forms will be molten so it can flow out into the mould ✓

d *(any two)* lustrous, good heat conductor, good electrical conductor ✓✓

4 **a** weight of copper $= \frac{34.6}{100} \times 500$ ✓

$= 173$ t ✓

b At each processing stage, not all the copper is recovered and so the mass of copper that is produced at the end will be less than the theoretical amount. ✓ (In addition, the ore is not pure chalcopyrite mineral and so the actual mass of copper it contains will be lower.)

c See the graph below.

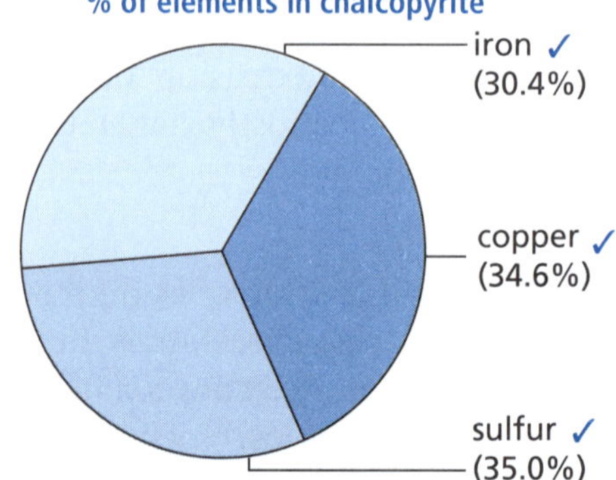

5 **a** % alumina $= \frac{550}{1000} \times \frac{100}{1}$ ✓
$= 55\%$ ✓
(note: 1 t = 1000 kg)

b mass of aluminium $= \frac{79.4}{100} \times 550$ ✓
$= 436.7$ kg ✓

6 **a** false ✓; **b** true ✓; **c** true ✓; **d** false ✓; **e** true ✓; **f** false ✓; **g** true ✓; **h** true ✓; **i** false ✓; **j** true ✓

MAINTAINING OUR LOCAL ENVIRONMENT
Ores and environmental issues Pages 114–115

1 *(any five)* less pressure on garbage tips (glass takes a long time to naturally break down in a landfill but recycled glass bottles can be back on shelves within a month or so); glass recycling is efficient (recycled glass is the primary ingredient in all new glass containers, containing up to 70% recycled glass); glass recycling is sustainable (they can be recycled repeatedly); glass recycling conserves natural resources (every tonne of glass recycled saves more than a tonne of the raw materials needed to create new glass); the process for glass recycling is simple; glass recycling saves energy (40% less energy is consumed than making new glass from raw materials); glass recycling pays (in many regions cash refunds are offered for most glass bottles) ✓✓✓✓✓

2 *(any four)* protect shores from wave action; reduce the impacts of floods; absorb pollutants and improve water quality; provide habitat for animals and plants; many contain a wide diversity of life, supporting plants and animals that are found nowhere else ✓✓✓✓

3 **a** Plants firmly rooted in the muddy bottom, but with stalks that rise high above the water surface, are able to radically slow the flow of water. ✓ Hence, they counter the erosive forces of moving water along lakes and rivers, and in rolling agricultural landscapes. ✓

b Wetlands prevent flooding by holding water much like a sponge. ✓ By doing so, wetlands help keep river levels normal and filter and purify the surface water. Wetlands accept water during storms and whenever water levels are high. When water levels are low, wetlands slowly release water. ✓

4 **a** per person ✓ (The term *per capita* is Latin and literally means 'by heads'.)

b around 300 kg ✓

c No, this is an average. ✓ Some people (businesses) used more, some less. It is hard to imagine each person used roughly 1 kg of paper each day.

d $290 \times 3\,000\,000 = 870\,000\,000$ kg ✓
$= 870\,000$ tonnes ✓
(1 t = 1000 kg)

e Extrapolating from the graph, this should be around 270 to 280 kg. ✓

f $\frac{90}{100} \times 330 = 297$ kg, so around 300 kg ✓

g There is very little evidence of this in the graph. ✓ Computers became popular in the last 20 years and there is very little evidence that this has led to a dramatic decrease in paper consumption. But then paper is used for a whole host of purposes: exercise books, newspapers, gift-wrapping and textbooks to name a few.

5 Paper is made mainly from wood fibre. The more paper we use, the more trees need to be cut down. ✓ (Over 80% of the world's ancient forests have already been cut down for fuel and for the raw materials that go into paper and other wood products.) Recycling paper reduces the amount of new or virgin wood needed for manufacturing by allowing discarded paper to be re-used. ✓ (The loss of forests causes huge damage to our planet. Trees store about 20% of the carbon dioxide on Earth, helping to stabilise greenhouse gas levels and our planet's fragile climate. Forests also provide habitats for huge varieties of plants and animals, and help to both preserve the cleanliness of our water supplies and prevent soil erosion.)

6 Nuclear accidents can spill radiation over a large area. ✓ This is especially so if particles are dispersed by air. Given that these radioactive particles can remain in the environment for decades and centuries (sometimes even millennia), areas will be unsafe for lengthy time periods. ✓ (The Chernobyl nuclear reactor explosion in the Ukraine in 1986 is the worst nuclear accident in history. An exclusion zone of 30 km remains in place today, although its shape has changed and its size has been expanded. Officials estimate the area will not be safe for human life again for another 20 000 years.) Oil spills, such as those from supertankers or oil rigs, can cover large areas and take a long time to clean up. ✓ Clean-up and recovery from an oil spill is difficult. Oil penetrates into the structure of the plumage of birds and the fur of mammals, eventually killing them. ✓ Animals also suffer digestive tract problems, altered liver function and kidney damage.

7 Plastics can remain in landfill areas for decades and centuries before decomposing, and so take up a lot of space. ✓ Plastics that find their way into marine environments can have harmful effects on animals. ✓ (Most of the marine debris in the world is plastic materials, between 60 to 80% of total marine debris.)

8 a 10% ✓

b 100 – (37 + 10 + 10 + 8 + 7 + 10) = 18% ✓

c Metals, paper and glass can be recycled, and so can most plastics. ✓ Some garden waste (leaves, branches) can be ground to mulch and re-used. ✓ The bars shown for these would be greatly reduced. ✓

9 a There has been a steady increase in both sales of aluminium cans and in recycling. ✓ However, recycling has not kept pace with the number of cans sold. ✓

b It takes the strain off our natural resources ✓, takes up less space in landfill ✓ and it is far cheaper to recycle aluminium cans than it is to produce new ones from the ore ✓. Making aluminium from bauxite ore is a dirty process. By recycling more, many tonnes of pollutants per year can be prevented from entering the atmosphere. (It takes 95% less energy to recycle old aluminium into new cans than to make aluminium from bauxite ore. In other words, 20 cans made from recycled aluminium using the same energy as just 1 can made from new aluminium!)

c a public advertising campaign pointing out the benefits of recycling ✓; 'cash for cans' program where collected cans can be taken to a recycling depot and exchanged for cash ✓

10 a *(accurately labelled axes ✓; vertical scale correct ✓; columns correct length ✓; graph title ✓)*

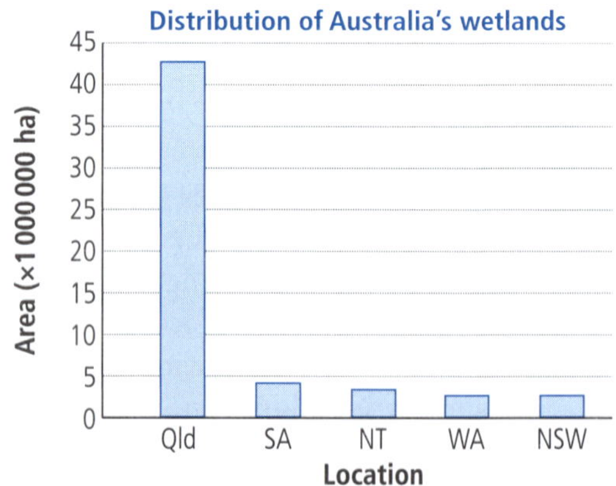

b At first glance this would appear to be true. ✓ After all, Queensland has at least 10 times more wetlands than any other state. However, you could also argue that wetlands are not distributed evenly across the country. ✓ There are many wetlands all over Queensland, which is also a large state. Victoria has 0.56 million ha set aside for wetlands, while Tasmania has only 52 000 ha. But then, these are smaller states. Tasmania has proportionally more wetlands for its size than any other Australian state. So you need to take into account the location of the wetlands and the area of the state/territory where they are located. This would be the approach taken by the federal government whereas as state governments would be biased towards their own states. ✓

SUSTAINABILITY

Ores and environmental issues Pages 120–121

1 **a** Crop rotation means rotating crops so that no bed or plot sees the same crop in successive seasons. ✓

b *(any two)* long-term improvement of soil quality and productivity; reduces the impact of pests and diseases (crop rotation disrupts weeds and insects and pathogen life cycles, making it difficult for pest and pathogen populations to build up); potential for less environmental impact (greater nutrient use and cycling, less use of pesticides and improved soil quality) ✓✓

c Legumes can dramatically reduce the need to purchase inorganic nitrogen fertilisers for crops. ✓ Through biological nitrogen fixation, most legumes increase the level of nitrogen in the soil. ✓ For example, alternating corn with alfalfa (or lucerne, a legume plant) can provide most, if not all, of the nitrogen needed by a subsequent corn crop.

2 **a** grow quickly putting on a lot of meat ✓, not just fat

b grow to a small size, but be a good egg layer ✓ (Growing to a small size means that the energy in feed is turned into eggs, rather than bulk up the chicken.)

3 There would be desirable features in both abalone species. These features can be combined together in one hybrid animal ✓ which, from the point of view of farmers and customers, would be superior to those in either native species ✓. Desirable features would include growth rate, survival, resistance to disease, meat yield and meat quality.

4 Fish with the desirable features are kept as the breeding population. ✓ Other fish are sold off. The fish with the desirable features are bred so that these features can be passed on to the next generation of fish. ✓

5 *(any five)* increased quality and yield of the crop; resistance to viruses, fungi and bacteria; increased tolerance to insect pests; increased tolerance of herbicides; being able to use water more efficiently (less strain on a farmer's water resources); increased tolerance to environmental pressures (such as salinity, extreme temperature and drought) ✓✓✓✓✓

6 Growth rate: A faster growth rate speeds up the turnover of production. ✓ Improved growth rates show that farmed animals and plants utilise their feed more efficiently.

Survival rate: This takes into account the degrees of resistance to diseases, as plants or animals under stress are highly vulnerable to diseases. ✓

Plant/animal matter: Quality is of great economic importance in the market and affects what customers are prepared to pay. ✓ Animal quality usually takes into account size, meatiness, percentage of fat, colour of flesh, taste and shape of the body.

Produce young: Selective breeding practices could ensure that young animals

or plants are produced relatively easily in sufficient numbers to grow the next generation. ✓

7 In this selective breeding program, the farmer is selecting for increased growth rate (by selecting for length) and is also selecting for weight. ✓ He has established independent cut-off values of 242 g for weight and 20 cm for the length of cob. Cobs which meet or exceed both cut-off values are saved and become the select plants to propagate for the next generation. ✓ All other corn cobs are sold.

8 Fire can create a rejuvenated feeding habitat for many animal species. ✓ Herbivorous mammals like the fresh pickings of new grasses and leaves associated with the regrowth following a fire. ✓ Succulent shoots can attract animals such as wallabies, wombats and grasshoppers.

9 **a** The thick woody trunks prevent the tree from being killed. ✓ While other plants around this tree may have died, new growth allows the tree to flourish in an environment without much competition from other plants, and due to an increase in access to sunlight for photosynthesis.

b The seed coat prevents the seed from being killed ✓ and therefore not surviving. Fire can liberate the seed and the young plants can then capitalise on the lack of competition ✓ in a burnt landscape.

c Being slow growing, they may not survive to maturity before the next fire rolls through. ✓ By having a protective base, the crown can regenerate and the plant continues its growth as before. ✓

10 Initially there is a range of lengths for the fish ✓, with a mean length of 100 mm. In each generation, the farmer keeps the longest fish from which to breed the next generation. ✓ Over a number of generations the average length of fish increases ✓, so he is eventually obtaining longer (and larger) fish. There is still a range of fish ✓ but now his shortest fish is the length of the average fish three generations earlier.

11 Underground tubers are food stores and out of direct damage from the fire, while the rest of the plant may have been burned down to ground level. ✓ Once the fire has passed, new shoots can emerge from these tubers. ✓ These plants, such as orchids, benefit from their increased access to light. Without fire, competition for sunlight leads to a marked decrease in their numbers.

Strand: Physical sciences

KINETIC AND POTENTIAL ENERGY
Energy Pages 126–127

1 **a** **i** Q ✓; **ii** P and R ✓
b **i** Y ✓; **ii** X and Z ✓; **iii** X ✓

2 **a** chemical potential energy ✓
b **i** margarine ✓
ii It contains large amounts of fat which stores more energy than carbohydrates and proteins. ✓
c An apple is mainly water and a small amount of carbohydrate. ✓
d *(one tick for each correct number)*

$$670 + 2\left(\frac{1160}{2}\right) + \frac{1}{4}(3000) + 2(250) + 1.5(290) = 3515 \text{ kJ}$$ ✓✓✓✓✓✓

3

	Example	Category
a	stretched rubber band	elastic PE ✓
b	E10 petrol	chemical PE ✓
c	water in a dam	gravitational PE ✓
d	snow avalanche	kinetic energy ✓
e	ocean wave	kinetic energy ✓
f	wound-up spring	elastic PE ✓
g	parachutist about to jump	gravitational PE ✓
h	compressed air in truck brakes	elastic PE ✓

4 KE = $\frac{1}{2}mv^2 = \frac{1}{2}(0.100)(20)^2$ ✓ = 20 J ✓

5 KE = $\frac{1}{2}mv^2$
450 = $\frac{1}{2}(1.0)(v^2)$ ✓
v^2 = 900
v = 30 m/s ✓

6 GPE = mgh
= 200 × 1.6 × 150 ✓
= 48 000 J ✓

7 **a** A = highest GPE as it is at the greatest height above the ground. ✓
B = lower GPE than at A as its height above the ground is less. Its GPE will only be one-quarter of its value at A as it is at one quarter of the height. ✓
C = GPE = 0 as the roller-coaster is now at ground level. ✓
b C ✓

ENERGY TRANSFORMATION
Energy Pages 131–132

1 **a** A = light energy into electrical energy ✓; B = chemical potential energy into electrical energy ✓
b The light meter reading will decrease ✓ as less light energy will reach the sensor as the distance of lamp increases ✓.
c *(Many answers are possible.)* change which metals are used; change the concentration of the acid mixture; change the temperature ✓✓

2 **a** the ears ✓
b the food being cooked ✓
c the ocean or land surface/trees/ buildings etc. ✓

3 **a** more energy ✓ The sound waves can be felt as the sound becomes louder, until the ear drum begins to hurt. ✓
b electricity supply or battery ✓
c more ✓

4 **a** heat energy into sound energy ✓
b The air particles in the tube begin to vibrate faster as they are heated and form a sound wave which moves up the tube and out into the air. ✓
c *(Answers will vary.)* alter the flame height by reducing the flow of gas to the flame to see if the sound volume decreases; change the diameter of the tube; alter the material of the tube ✓

5 *(Other answers are possible.)*
a burning petrol ✓
b moving a wire coil in a magnetic field to produce electrical energy ✓
c photosynthesis in plants ✓
d playing an electric guitar ✓
e dynamo on a bike to produce electricity to power the light ✓

6 **a** false ✓; **b** true ✓; **c** true ✓; **d** false ✓; **e** false ✓

7 *(Many answers are possible.)*
a solar cell ✓
b fan ✓
c television ✓
d burning coal to heat water to steam ✓
e a lift ✓

8 See the flowchart below.

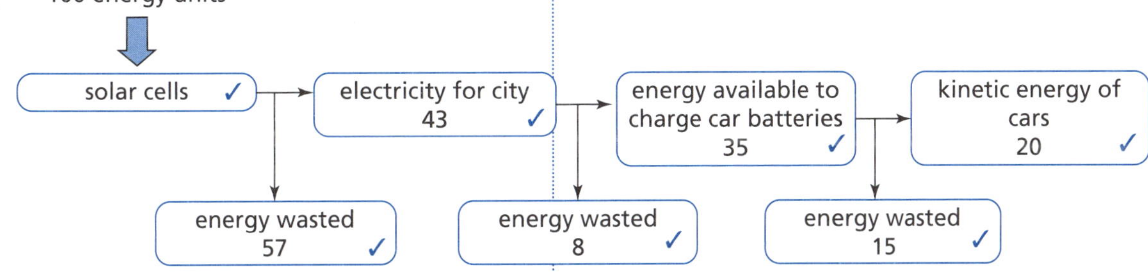

REVISION TESTS
Answers

CHECK YOUR ANSWERS

HEAT ENERGY
Energy Pages 136–137

1 **a** Heat is the total energy of molecular motion in a substance ✓ while temperature is a measure of the average energy of molecular motion ✓ in a substance.

b Temperature is measured in degrees Celsius ✓, while heat is measured in joules ✓.

c *(1 mark for each correct answer)*

This property depends on the ...	Heat content	Temperature
speed of the particles.	yes ✓	yes ✓
number of particles.	yes ✓	no ✓
size or mass of the particles.	yes ✓	no ✓
type of particles.	yes ✓	no ✓

2 P = convection ✓; Q = conduction ✓; R = radiation ✓

3 **a** The water at the top of the test tube is made to boil yet at the bottom of the test tube only a few centimetres away, ice remains unmelted. ✓ This shows poor heat conduction from the top of the test tube to the bottom. ✓

b The copper strip is heated in the middle so heat is travelling outwards from this point. So both the water and the mercury obtain the same amount of heat. ✓ That the glass bead falls from the mercury test tube shows that heat has travelled better through the mercury to melt the wax. ✓

4 Jenna is wrong. It is heat that will increase or decrease the temperature. If heat is added, the temperature increases. If heat is removed the temperature drops. Heat is transferred spontaneously from objects of higher temperature to ones of lower temperature (warmer to colder bodies). ✓ A fridge doesn't pump cold in, rather it pumps heat out. ✓ It is able to take heat from food and eliminate it out the back of the fridge.

5 **a** 32 °F ✓; **b** 37 °C ✓

6 Metals feel colder because, being good heat conductors, they conduct heat away from your hand. ✓ You perceive the heat that is leaving your hand as cold. ✓ (Believe it or not, the metal end of the screwdriver is not colder! It just feels that way.)

7 **a** Feathers, fur and clothing have air pockets and this air is a poor thermal conductor. ✓ Air pockets aid in cutting back on the heat loss through the material. ✓ (Clothes keep us warm not because they are warm, but because they prevent us losing body heat to the surroundings.)

b Styrofoam cups are poor conductors of heat because of all the air pockets they contain. ✓ In general, solids and liquids are very good conductors of thermal energy. Gases, however, are not. When gases are injected into the liquid polystyrene, they become trapped inside the resulting foam, forming little bubbles. ✓

8 *(take ½ mark off for each out of sequence)* C, I, E, H, A, G, D, B, F ✓✓✓

9 Steam and water can both exist at 100 °C. A steam burn is worse than a hot water burn because the steam is in a different phase. When the steam comes in contact with skin, the steam must first condense into liquid water before it can cool down to body temperature. ✓ This releases more energy into the skin and so causes a worse burn.

ENERGY EFFICIENCY
Energy Pages 141–143

1 Values missing from the table: 122.4° ✓, 86.4° ✓, 18° ✓, 7.2° ✓. See the graph on the next page.

(1 mark for correctly measuring angles, 1 mark for correctly labelling, 1 mark for title)

How energy is used in the home ✓

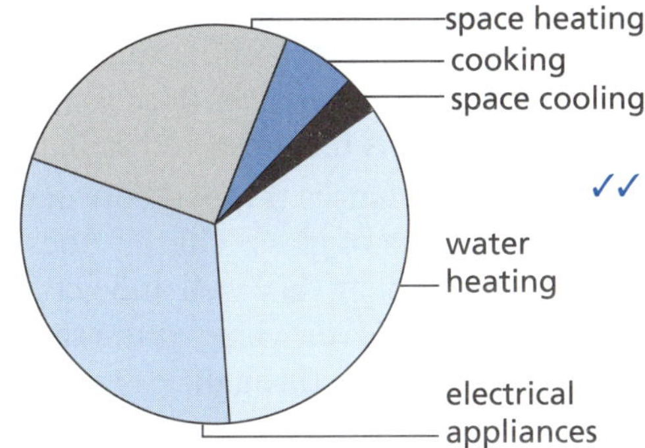

2 **a** Much heat in a room can be lost through the ceiling as warm air rises. This prevents heat loss through the ceiling. ✓

b Glass is not a good insulator so heat can be lost through a thin window. Double (or triple) glazing traps a layer of air between panes of glass preventing heat loss. ✓

c Blinds and curtains are not good heat conductors. They pose a barrier between the window and the room. ✓

d The higher the star energy rating for an appliance, the more efficient it is. That is, the more stars, the less energy it uses. ✓

3 **a** Stand-by power is the energy used by an appliance when it is not performing its main function. Many appliances use stand-by power to maintain an internal clock or to receive remote control signals. ✓

b Any three devices will do; examples include: television and home entertainment systems; computers; cordless telephones and fax machines; microwave ovens; battery chargers for portable devices such as mobile phones; any appliance with a remote control; any appliance with a built-in electronic clock. ✓

c Ensure appliances are not using stand-by power by switching them off at the wall. ✓ Use power boards with switches to make it easier to turn several appliances off at once.

4 **a** The average number of litres of petrol used for each 100 km travelled.

b If 10 L/100 km yields 3450 kg CO_2, then each 1 L/100 km yields 345 kg CO_2.
So to complete the column, the values downwards are: $6 \times 345 = 2070$ ✓; $8 \times 345 = 2760$ ✓; $12 \times 345 = 4140$. ✓

c The amount of petrol used will depend on how far is travelled. So the first asterisk (*) indicates that these figures are based on an average distance of 15 000 km. Similarly, how much this petrol costs will depend on the price at the bowsers, which changes over time, and which petrol station you go to. So the cost in this column is based on an average of \$1.60/L. ✓

d If 15 000 km is travelled, this represents 150×100 km. Since every 100 km uses 6 L, then the amount of petrol used is $6 \times 150 = 900$ L. ✓

e In order to complete the column, the values downwards are: $8 \times 150 = 1200$ ✓; $10 \times 150 = 1500$ ✓; $12 \times 150 = 1800$ ✓.

f 900 L @ \$1.60/L $= 900 \times 1.60$
$= \$1440$ ✓

g In order to complete the column, the values downwards are:
$1200 \times 1.60 = \$1920$ ✓;
$1500 \times 1.60 = \$2400$ ✓;
$1800 \times 1.60 = \$2880$. ✓

h The lower the fuel consumption of the car, the more savings can be made in dollar terms ✓, and also in CO_2 emissions.

5 **a** A hybrid vehicle uses two or more distinct power sources to move it. ✓ These combine an internal combustion engine (using petrol) and one or more electric motors (using electricity). ✓

b With increasing petrol costs, people are looking to cheaper ways to power their cars. Electricity is a cheaper form of energy than petrol but a back-up petrol engine is available should the electrical power run out. ✓ And, as technology improves, the range travelled by these vehicles before needing the batteries recharged is increasing.

c The batteries of the vehicle can be plugged in to house mains electricity for charging ✓, as well being charged while the engine is running.

d LPG (Liquefied Petroleum Gas) is used as well as biofuels such as ethanol. ✓

6 **a** Lighter cars use less fuel because they need less energy to start and stop than heavier cars. ✓ (Remember: kinetic energy depends on both mass and velocity, so you can obtain the same velocity with less energy if the mass of the vehicle is lighter.)

b *(any three)* a number of features including cost competitive (no more expensive than current car parts); perform equally well; durable (they last at least as long as steel parts); at least as strong ✓

7 **a** Solar radiation enters an enclosed insulated container where it is changed to heat. The air gap between the plate and sheet traps this heat, preventing it from escaping back into the atmosphere. ✓ Heat is transferred to the water circulating in tubes within the collector. Hot water exits at the top and is stored in a tank until it is needed.

b As water is heated it expands and rises ✓, creating a natural flow through the system.

8 **a** A thin pane of glass is not very good at preventing heat escaping from a room. ✓ Thicker and opaque walls are much better. (We put up with their mediocre thermal performance because we want the daylight, views and ventilation they offer.)

b Windows allow heat transfer by conduction across the glass ✓ and frames. And they allow sunlight to radiate through ✓, both as visible light and as invisible, infra-red radiation.

c A layer of air is trapped between two sheets of glass. This air does not transmit heat very well and acts as an insulating layer while, at the same time, allowing light through. ✓

9 **a** A solar cell converts photons of sunlight (very small packets of electromagnetic radiation energy) into an electrical current ✓ that can be used to power electrical loads.

b Living in a location that is not serviced by the main electric utility grid (off-grid living) makes it difficult and expensive to install electricity poles and cabling from the nearest main grid access point. ✓ A solar electric system is potentially less expensive.

c Electricity based on the Sun is renewable (it is always present), and does not rely on burning fossil fuels (such as coal or oil) which can generate pollutants and CO_2, which is a greenhouse gas. Hence it is clean. ✓

d *(½ mark for each)* solar panels have no moving parts and require little maintenance ✓; the electricity it produces for the system's lifespan is free ✓

e *(½ mark for each)* can have high initial installation costs ✓; not suitable for all locations ✓ (think of areas which don't get much sunshine)

Skills

FAIR TESTING
Investigations and problem solving Pages 148–149

1 **a** This is scientific as it can be easily tested by experiments. ✓

b This is scientific as it can be easily tested by experiments. ✓

c This is unscientific as it is a matter of personal taste. ✓

d This is unscientific as there is no standard test that has been given to Australians of the same age and educational level in this time; intelligence is difficult to measure as it is not just one's ability to remember facts. ✓

e This is scientific as it can be easily tested by experiments. ✓

2 **a** size, colour and shape ✓

b **i** size and shape ✓

ii colour ✓

iii temperature ✓

iv 1. Select three different coloured pieces of rubber foam that are all the same shape (e.g. cubes) and size (10 cm sides). ✓

2. Place glass thermometers in a small cut in each foam piece so the bulb is at the centre. ✓

3. Place the apparatus on a table in a sunny position in the yard. Measure the initial temperature. ✓

4. Measure the temperature of each piece after one hour. ✓

3 **a** A qualitative experiment is one in which observations are recorded but the quantities (e.g. mass, volume) are not. ✓

b **i** The yeast have died at this temperature. ✓

ii The yeast in the 40 °C experiment will not ferment sugar if the temperature of the tube is reduced back to 20 °C. ✓

iii No gas bubbles will be released. ✓

c **i** 1. Place the sugar and yeast mixture in a side-arm test tube that is clamped in the water bath, which is set at a fixed temperature (say 20 °C). ✓

2. Fill the measuring cylinder with water and invert it in a beaker of water. Place the end of the rubber hose at the mouth of the measuring cylinder. ✓

3. Collect the gas in the measuring cylinder for a fixed amount of time (e.g. 5 minutes). Measure this gas volume. ✓

4. Repeat the experiment a total of six times at different temperatures. ✓

ii temperature ✓

iii volume of gas ✓

iv *(Other answers are possible.)* mass of sugar ✓; mass of yeast ✓; same sized test tube ✓

v In a quantitative experiment, quantities such as mass and volume are measured whereas in a qualitative experiment only relative terms such as faster or slower are used. ✓

d **i** See the graph below.

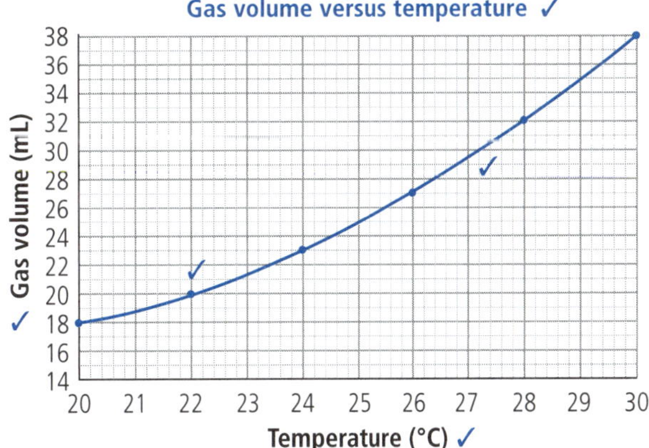

ii The rate of gas production during fermentation of sugar increases as the temperature increases up to 38 °C. ✓ (Note: the yeast die at 40 °C.)

4 **a** true ✓; **b** true ✓; **c** false ✓; **d** false ✓; **e** false ✓

REVISION TESTS
Answers

CHECK YOUR ANSWERS

WORKING IN THE LABORATORY
Investigations and problem solving Pages 154–155

1 **a** The rod acts a guide for the acid and reduces the risk of the acid splashing onto your skin or into your face. ✓
b If splashing occurs then safety glasses will protect your eyes. ✓
c **i** No ✓ The acid is not flammable but students should exercise great care when heating with a Bunsen flame. ✓
ii No ✓ Blue flames are hard to see, especially in a sunny room. The flame should be yellow for safety. ✓

2 **a** The beaker should never be filled to the top as it will overflow when heated. ✓
b A burning scrap of paper is hard to control and can burn your hand. A match or taper should be used. ✓
c A closed hole produces a yellow flame which is sooty and will blacken the apparatus. This is a poor experimental method. ✓
d Students must monitor the experiment at all times in case an emergency arises. ✓

3 **a** false ✓; **b** true ✓; **c** false ✓; **d** false ✓; **e** true ✓

4 **a** See the diagram below.

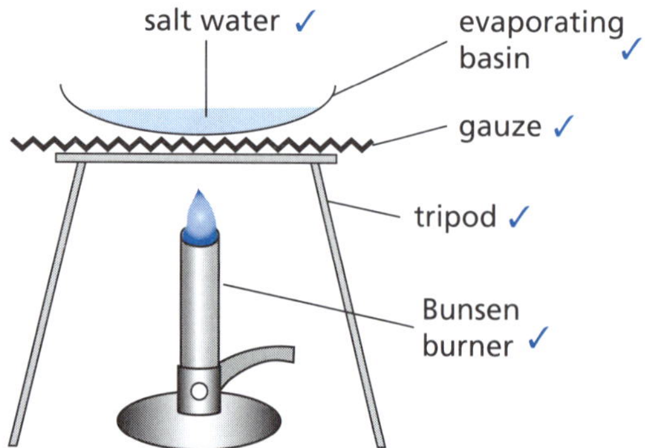

b The apparatus will get very hot and so it should not be touched until the Bunsen is turned off and it is allowed to cool down. ✓ Heating should be gentle so the salt solution does not bubble and boil dangerously. ✓ Safety glasses and a lab coat should be worn to protect the eyes and body. ✓

5 A good answer should include the following points.
- The type of cage to be used. The size, materials and ease of cleaning should be considered. ✓
- Food and water containers. How often do the mice need to be fed and provided with fresh water? ✓
- Floor litter and nesting materials. What is the best litter material and how often does it need to be changed? Mice prefer shredded paper for nesting. How often does this need to be changed? ✓

6 **a** Hydrochloric acid is corrosive and can attack the skin or eyes. Wear safety glasses to protect the eyes and a lab coat to protect the body. ✓
b Hydrogen is explosive in air and so to avoid the possibility of an explosion, there should not be any naked flames nearby. ✓

FIRST- AND SECOND-HAND INVESTIGATIONS
Investigations and problem solving Pages 159–160

1 **a** Single blind ✓, as Mary knows which biscuit is made with cocoa and which with chocolate ✓, but her friends don't ✓.
b The other ingredients she uses must be kept the same ✓ (the same amount of flour, sugar, etc). Also, the biscuits need to look the same and be the same size and shape. ✓
c No, some people may have a preference for cocoa and others for chocolate so she needs a number of friends ✓ to help make the distinction clearer ✓.

2 In a single-blind test with the witness there may be subtle or overt influence ✓

by the officer if they know who the suspect is. It might not be intentional; it might simply be body language. ✓ Defence lawyers could pick up on this and ask that the evidence not be admitted.

3 **a** Volunteer test subjects (patients) are randomly assigned to experimental and control groups. ✓ They receive identical treatment ✓ but neither the patient nor the doctor prescribing the treatment knows whether they are getting the experimental drug or the placebo ✓. The treatment is allowed to run its course and the results from the two groups are compared. ✓

b Each patient may experience slightly different results. ✓ Some patients may show side effects, others may not. Some may be allergic to the drug, others may not. Still some patients might recover faster than others. So it is necessary to have a number of subjects (in some experiments this can run into hundreds, if not thousands) to help even out those differences. ✓

c They should be of similar composition ✓: age, gender, severity of the condition, ✓ and so on. Otherwise, these factors might come into play. For example, young people might be expected to make a speedier recovery than older people, regardless of the effect of the drug.

d This might consciously or subconsciously affect the results. ✓ There is strong evidence of a placebo effect with medicine. This is where, if people believe that they are receiving the active drug, they may show some signs of improvement in health. A blind experiment reduces the risk of bias from this effect, giving an honest baseline for the research and allowing a realistic comparison. ✓

e While the vast majority of researchers are professionals, there is the possibility that the researcher might subconsciously communicate information to a patient about the pill they are receiving. ✓ Some might even favour giving the pill to patients that they thought had the best chance of recovery, hence skewing the results. Also, there is often pressure from billion-dollar drug companies to generate positive results. ✓ And there is always a chance that a scientist might manipulate the results, putting a positive spin on them. A double-blind experiment reduces the chance of this kind of criticism and minimises the effects of subjective judgements. ✓

f *(one mark for any reason)* Suppose the procedure was used in conjunction with surgery. As there are risks with surgery, and no possible benefit to the patient taking the placebo, this poses an ethical dilemma. Or suppose the experimental drug makes the patient feel nauseous (sick) and the placebo does not, the patient can tell which drug they are receiving. It would then be unethical to put some nauseating chemical in the placebo. Also, is it morally correct to purposely not treat the placebo group of participants who may be seriously or terminally ill? ✓

4 **a** Access to information is far easier today, especially with the availability of the Internet. ✓ So the researcher only needs to locate the source of the information. The bulk of secondary research data gathering does not require using expensive and specialised equipment ✓, nor highly trained personnel to gather it ✓. Some research can be expensive to perform, but secondary research expenses are paid for by the originator of the information.

b Before you can obtain the secondary data you are seeking, someone needed to have this information published.

This can take time. ✓ Often journals may take up to 6 months before the information appears in print, and this could be for research that was carried out perhaps a year or more earlier. ✓ Self-publishing on the Internet is faster. But there is also the problem of how long this information has been around before you come across it. ✓ And has this older information, still in print or on the Internet, been updated with more current information? ✓

c People often wanting to push their point of view will often self-publish the information or have it appear in documents or websites where other people hold similar views. ✓ For example, there are sites where some people still believe the world is flat, where the Moon landing was faked in the back lots of some movie studio, where immunising children against dangerous diseases is harmful and where unassisted home births are safe. These four positions are not held by mainstream science. It is therefore important to check out the source of this information and to compare it with that given at other sites. ✓

5 Members of a team have different personalities and backgrounds. Each has their own individual set of strengths and weaknesses. ✓ Understanding your team colleagues can be the answer to building good working relationships. A team with too many similar people may enjoy working together, but find that they struggle to meet deadlines for important tasks. In other words, they might not have the scope of knowledge and abilities needed to complete some project. ✓ Many questions in science today are complex and require researchers of different backgrounds and skill sets to help answer them.

6 Scientists often work in teams, so it's good to develop these skills of working scientifically ✓; teams enhance learning effectively from each other ✓; teams are more effective than individuals when working on complex projects ✓; teamwork develops interpersonal skills ✓.

7 **a** primary ✓; **b** secondary ✓;
c secondary ✓; **d** primary ✓

EXPERIMENTAL REPORTS
Investigations and problem solving Pages 165–166

1 **a** **i** mass of fuel P ✓
ii altitude reached ✓
iii mass of fuel Q ✓

b Repeat trial numbers 3 to 5, but making smaller fuel P increments. ✓ This way the scientist can determine which mix allows the rocket to reach maximum height.

c It is important to keep all the other variables constant. ✓ Different rockets can travel different distances on the same amount of fuel mix for a variety of other factors. These all need to be kept constant.

d If more than one variable changed at a time then the scientist couldn't be certain which of these was responsible for the maximum height reached by the rocket. ✓

2 The amount of fertiliser is the variable factor, and the use of two similar plants in the same location receiving equal amounts of water implies the notion of a control. However, with such a small sample size (only two plants) and no repetition ✓, this is a flawed experiment. Plants of the same species can respond differently. What if one of the plants died for some unrelated reason, such as the trauma of being re-planted? Would she then erroneously put this down to the fertiliser or not the fertiliser? Having a number of plants undergoing identical treatment will even out slight variations due to a particular plant. ✓

3 **a** to determine the effect of different salt concentrations on the health of *Elodea* ✓

b to serve as the control ✓ This is the base line to determine how the other plants are faring. ✓

c increasing amounts of salt concentration ✓ Solutions can vary but a possible set for the other four test tubes could be: pond water + ½ teaspoon salt per 100 mL solution; pond water + 1 teaspoon salt per 100 mL solution; pond water + 1½ teaspoon salt per 100 mL solution; pond water + 2 teaspoon salt per 100 mL solution. ✓ If there is no effect, then the salt concentrations can be increased.

d the healthiest-looking plants ✓ She could record whether the plant looks healthy, is growing or whether the leaves are yellowing, or the stems are becoming limp.

e in the form of a table ✓ The following is one example.

Test tube	#1 (control)	#2	#3	#4	#5
Day 1					
Day 2					
Day 3					

4 **a** to determine the effect of the distance a light source is placed from *Elodea* on its production of oxygen ✓

b This is one simple way to vary the light intensity falling on *Elodea*. ✓ The further away the lamp is, the lower the light intensity is. (Another way is to keep the distance the same but have a dimmer control inserted with the lamp.)

c Submerge a specimen of *Elodea* pondweed in a large beaker of water. ✓ Cover it with an inverted funnel. ✓ Fill a test tube completely with water and place this upside down on top of the stem of the funnel. ✓ Place the lamp a given distance away from the beaker. ✓ Shine a strong light source on the plant. ✓

d One of the controlled variables needs to be temperature and this needs to be kept constant throughout the experiment. ✓ If the temperature rises this could affect the rate of oxygen production or the health of the plant ✓ and this would then not be a fair experiment.

e Rate of oxygen production in *Elodea* with varying light intensities. ✓

f The further away the light source is placed, the fewer bubbles are seen per minute. ✓

g the size of the *Elodea* specimen ✓; the temperature of the water ✓

h Instead of an inverted test tube, use an inverted measuring cylinder. ✓ That way you can measure the volume of gas ✓ produced and not just count bubbles. This would make the experiment easier ✓ as you would only need to read the volume on the cylinder after a given time period. In other words, you don't need to count bubbles. Counting so many bubbles can lead to errors as you might miss some or double-count. ✓ Another problem is that bubbles are not all the same size.

SAMPLE EXAM PAPER 1 Pages 169–173

Part A: Multiple-choice questions

1 **B** ✓ The cytoplasm is the gel-like substance inside the cell holding all the cell's internal substructures (called organelles). Vacuoles are present in all plant and fungal cells and some protozoa, animal and bacterial cells, so A is incorrect. Ribosomes and the nucleus are not jelly-like, so C and D are incorrect.

2 **A** ✓ Digestion begins in the mouth. Chewing grinds food into fine particles, which are moistened and mixed with saliva containing the enzyme ptyalin. Ptyalin changes some of the starches in the food to sugar. So B, C and D are incorrect.

3 **D** ✓ Messages are sent along nerves causing muscles to contract. A is incorrect as this has to do with the heart and blood circulation. B is incorrect as muscles don't move themselves. C is incorrect as the skeleton moves as a result of muscles pulling on them.

4 **B** ✓ A is wrong as each cell contains the same number of chromosomes as the parent cell. C is incorrect as daughter cells are identical to the parent cell. D is wrong as the cell division is mitosis.

5 **C** ✓ DNA, or deoxyribonucleic acid, is the hereditary material in humans and almost all other organisms. It replicates within the nucleus. Hence A, B and D are incorrect.

6 **B** ✓ This is part of the urinary system, while the others (A, C and D) are all part of the digestive system.

7 **D** ✓ All statements are correct.

8 **B** ✓ Fertilisation, also known as conception, is the fusion of gametes to produce a new organism. A, C and D are incorrect as only B refers to the physical union of male and female gametes, of sperm and ova in an animal or pollen and ovule in a plant.

9 **D** ✓ Molecules vary in size, mass and how many atoms make it up, hence A, B and C are wrong. All particles are in random motion, making D correct.

10 **B** ✓ The trachea is the windpipe. The oesophagus is the food pipe, so A is incorrect. Alveoli are the final branching of the respiratory tree and act as the primary gas exchange units of the lung. As alveoli are in the lungs, C is wrong. The pulmonary artery brings blood to the lungs to be oxygenated, and to release carbon dioxide. Hence D is wrong.

11 **C** ✓ In order to form a Ca^{2+} ion, a calcium atom gives up two electrons, so B is incorrect. The numbers of protons and neutrons do not alter, hence A and D are wrong.

12 **D** ✓ Blood plasma is the straw-coloured liquid component of blood, making up about half of the total blood volume, and is around 90% water. Without plasma, blood cells would have no medium to travel in as they moved through the body. It delivers various substances to the cells of the body, and collects waste products for processing. A, B and C are wrong as these all travel within the plasma.

13 **B** ✓ The egg cell contains a reserve of nutrients which fuels the initial cell divisions and growth in the early development of the embryo. A is incorrect, as an egg cell is only one cell. C and D are incorrect as, on average, these are no larger than those in sperm cells.

14 **A** ✓ House dust mites are a common cause of asthma and other allergic symptoms. The mite's gut contains powerful digestive enzymes that endure in their faeces and can induce allergic reactions such as wheezing. Its exoskeleton can also contribute to allergic reactions. Dust mites are arthropods that belong to the arachnid class, the same class to which spiders belong. Spiders are not insects, while flies and wasps are. Hence B, C and D are incorrect.

15 **A** ✓ Bile is a bitter-tasting, dark green to yellowish brown fluid, produced by the liver of most vertebrates, which aids the process of digestion of fats and oils in the small intestine. In many species, bile is stored in the gall bladder and released into the duodenum, the first part of the small

intestine just after the stomach. Hence B, C and D are incorrect.

Part B: Restricted-response questions

16 mitochondrion ✓

17 arteries ✓

18 spore ✓

19 cell wall ✓ (Only the green parts of plants contain chloroplasts to carry out photosynthesis. Plant roots, for example, do not have chloroplasts.)

20 leucocytes ✓

21 virus ✓

22 gaseous ✓

23 gastric ✓

24 bacteria ✓

25 ovary ✓

Part C: Written knowledge and skill questions

26 Early observers noted the wriggling sperm and thought they saw a miniature, fully formed human body located inside the head of the sperm. The homunculus theory argues that the father provides all the characteristics while the mother is only the growing and feeding receptacle. In other words, the mother is a convenient container in which the male-generated miniature man or woman could grow and be nurtured. This would be an example of asexual reproduction. Today we know that the sperm contains 50% of the chromosomes of the new baby with the mother's egg contributing the other 50%. ✓ Humans reproduce sexually. ✓

27 **a** melting an ice cube ✓

b A chemical change produces a new substance. Physical changes are concerned with energy and states of matter. A physical change does not produce a new substance. ✓

c *(any two)* observations such as light, heat, colour change, gas production, odour or sound ✓✓

28 **a** true ✓; **b** false ✓ (magnification is 100×);

c true ✓

29 **a** covalent ✓

b An ionic bond, generally the bonding between a non-metal and a metal, occurs when charged atoms (ions) attract each other after one loses one or more of its electrons to the other ion. ✓ It is the attraction between two oppositely charged ions. A covalent bond is a chemical bond that is displayed by the sharing of pairs of electrons between atoms. ✓

c covalent ✓ Plastics are actually polymers of simple units joined by covalent linkages.

30 **a** *(any two)* good conductor of heat and electricity; malleable; ductile; solid at room temperature (except mercury); metals can combine with other metals (and some non-metallic elements) to form a vast number of alloys; lose their valence electrons (electrons in the outermost shell) easily; possess metallic lustre; opaque as a thin sheet ✓✓

b **i** zinc oxide ✓

ii magnesium chloride ✓

iii oxygen ✓

31 **a** platinum ✓

b 11 ✓

c This is a molecular model of a chemical substance which presents both the three-dimensional position of the atoms and the bonds between them. The atoms are normally represented by spheres, connected by rods (sticks) representing the bonds. ✓ This is a visual representation as molecules don't look like this.

32 Flowering plants have both male and female reproductive organs. The stamen is the male reproductive organ. Pollen grains are found on the anther. ✓ The carpel is the female reproductive organ and is made up of the stigma, style and ovary. ✓ Pollen can land on the stigma, having been transferred there by the wind or insects (pollination). A pollen tube develops down the style so the male nucleus can combine with a female nucleus in the ovary to produce a seed. ✓

SAMPLE EXAM PAPER 2

Pages 174–179

Part A: Multiple-choice questions

1 **A** ✓ B is incorrect as it refers to the coarse focus knob. C is incorrect as it refers to the stage of the microscope. D incorrect as this refers to the iris diaphragm.

2 **A** ✓ B is wrong as veins have thinner walls than arteries. C is wrong as the blood pressure is lower in veins. D is wrong as gaseous exchange occurs via capillaries.

3 **B** ✓ A is wrong as ethanol is a molecular compound. C is wrong as ethanol is a pure substance. D is wrong as the ratio is 1:3.

4 **C** ✓ A is wrong as chemical changes leads to new substances. B is wrong as chemical changes are difficult to reverse. D is wrong as condensation is a physical change.

5 **C** ✓ A, B and D are wrong as each is too reactive to exist as a native metal.

6 **D** ✓ A is wrong as froth flotation just separates the mineral from the unwanted rock. B is wrong as electrolysis is used to purify the copper. C is wrong as this is not used at all.

7 **A** ✓ B is wrong as this refers to crop rotation. C is wrong as this refers to resources that should not be removed if they can't be replenished. D is wrong as it refers to burning off the dry undergrowth in forests.

8 **B** ✓ A is wrong as kangaroos and koalas do not live in wetlands. C is wrong as dams and reservoirs are used for this purpose. D is wrong as wombats and possums do not live in wetlands.

9 **B** ✓ A, C and D are wrong as these are only minor or inconsequential reasons.

10 **C** ✓ A is wrong as sand is used for this. B is wrong as you should evacuate if there is a fire. D is wrong as the blanket is not absorbent.

11 **B** ✓ A, C and D do not involve nuclear processes.

12 **D** ✓ A is wrong as food and fuels store chemical potential energy. B is wrong as this refers to position above the ground. C is wrong as this is the energy of motion.

13 **B** ✓ A is wrong as granite is intrusive. C and D are wrong as they have low silica contents.

14 **A** ✓ B is wrong as gneiss is formed from granite. C is wrong as slate is formed from shale. D is wrong as marble is formed from limestone.

15 **C** ✓ A is wrong as a tooth is hard and readily fossilised. B is wrong as this is a whole-animal fossil. D is wrong as this is wood that has been mineralised and turned to stone.

Part B: Restricted-response questions

16 membrane ✓

17 pollinated ✓

18 diffusion ✓

19 continents ✓

20 expands ✓

21 stored ✓

22 transformation ✓

23 magma ✓

24 explanation ✓

25 cold ✓

Part C: Written knowledge and skill questions

26 **a** contact metamorphism ✓

b X = shale ✓; Y = quartzite ✓; Z = marble ✓

c sedimentary ✓

d Y has larger, coarser crystals. ✓

27 **a** The term literally means 'hot water'. Hydrothermal deposits refer to minerals that have been dissolved and crystallised from hot concentrated solutions in volcanic areas. ✓

b **i** A mineral vein or lode is formed when minerals crystallise from cooling, concentrated solutions. ✓

ii Chemical weathering of chalcopyrite leads to the formation of malachite and azurite. ✓

c *(½ mark for each)* A azurite = iii weathered mineral; B silica = iv slag; ✓ C smelting = i using a high-temperature furnace; D magmatic heat = ii hot water leaching ✓

28 **a** elastic potential energy ✓

b elastic potential energy to kinetic energy ✓

c to make the car as light as possible as the balloon does not store sufficient energy to drive forward a heavy car ✓ (A lot of energy would be wasted in overcoming friction against the floor if the car was heavy.)

29 **a** AB and CD ✓

b BC ✓

c GPE decrease = mgh A – mgh B
= 686 000 – 294 000
= 392 000 J ✓
alternatively, 500 × 9.8 × (140 – 60)
= 392 000 J

d 392 000 J ✓

e KE of C = GPE lost from A to C
= 500 × 9.8 × 40
= 196 000 J ✓

30 **a** dependent ✓

b 1. Prepare bread dough rolls of the same mass and size (circumference). Keep in the fridge until used. ✓
2. Use five rolls in each experiment as this will increase reliability. ✓
3. Perform the experiment three times under three different temperature conditions (e.g. 15 °C, 25 °C and 35 °C) keeping the time for leavening constant for each experiment (e.g. 1 hour). ✓

31 The soft parts decay away leaving the shells. The shells are crushed and compressed under layers of overlying sediments. ✓ The powdered minerals (calcium carbonate) then become cemented together to form limestone rock. ✓

TEST & EXAM RESULTS

TEST RESULTS

Transfer your **Percentage Score** that you calculated in the **Your Feedback** box at the end of each **test** and **exam** to the table below. This will help you work out your areas of strength and weakness.

Topic	Percentage Score
Microscopes	%
Cell structure	%
Cell division and reproduction	%
Disease	%
Multicellular organisms	%
Digestive system	%
Circulatory and respiratory systems	%
Excretory and skeletomuscular systems	%
Asexual and sexual reproduction	%
Biotechnology	%
Matter and changes of state	%
Elements	%
Compounds	%
Simple chemical reactions	%
Earth structure and minerals	%
Weathering and erosion	%
Sedimentary rocks and fossils	%
Igneous and metamorphic rocks	%
Metals and ores	%
Extracting metals from common ores	%
Maintaining our local environment	%
Sustainability	%
Kinetic and potential energy	%
Energy transformation	%
Heat energy	%
Energy efficiency	%
Fair testing	%
Working in the laboratory	%
First- and second-hand investigations	%
Experimental reports	%
Sample Exam Papers	
Paper 1	%
Paper 2	%

FEEDBACK CHECKLIST TO IMPROVE YOUR TEST & EXAM RESULTS

Do you want to improve your scores in the Revision Tests and the Sample Exam Papers? Check that:

You are ready to do a test or exam.

- You need to revise your class notes.
- You need to revise the relevant sections of your school textbook.
- Do further revision using the ***Excel*** *Science Study Guide* (available for Years 7, 8, 9 and 10).
- Use the Quick Revision and Revision Summaries in this book to prepare yourself for each Revision Test and Sample Exam Paper.

Your standards are realistic.

- Use these Revision Tests and Sample Exam Papers to determine your strengths and weaknesses.
- If you make mistakes then you need to learn from them. This is the ultimate key to success.
- Once you identify any weaknesses, spend more time trying to improve your understanding in these areas by doing more revision using your notes, textbook and science study guide.

You have planned your answers.

- Do not rush your responses to each question. Spend a little time thinking about the best way to construct a good reply.
- For mathematical questions, show all your working. Once you have completed your calculation, double-check to make sure that you have not made any errors.
- For written questions, present your well-planned responses in a logical order. Sub-headings may be useful for longer responses.
- Present all relevant information, even if you think some of this is common sense.

Your answer correctly responds to the verbs used in the question.

- In science, the verbs used in the question require different types of responses.
- The verb *identify* requires you to provide a name for a thing or process.
- The verb *explain* requires you to relate cause and effect, and to show the relationships between things.
- The verb *describe* requires you to provide characteristics and features of an object or living thing.
- The verb *discuss* requires you to identify issues and provide points for and/or against.
- The verb *predict* requires you to suggest what may happen based on available information.
- The verb *account* requires you to state reasons for an event or process.
- The verb *evaluate*, in a mathematical sense, means 'calculate or find the value of'.

INDEX

A

accidents 152
acid rain 84
active immunity 20
agriculture 118–119
aim of investigation 162
algae 19
alimentary canal 30–31
aluminium 106–107
alveoli 36
analysis section 164
angiosperms *see* flowering plants
animals
  cells in 8
  cloning of 52
  experiments using 153
  farming 118
  fossils of 90
animal pollinators 47
antibiotics 19, 20
antibodies 20
antigens 20
aorta 35
arteries 35
artificial insemination (AI) 51
artificial selection 118–119
asexual reproduction 45–49, 51
assisted reproductive technology 51–52
atoms 57
  in chemical reactions 74
  in compounds 68
  in elements 62
  in minerals 78
atomic number (Z) 62
atomic weight 62
Australia
  bushfires in 119
  mining sites in 106
  wetlands in 111–113

B

bacteria 19, 46
ball and socket joint 40
bauxite 106–107
bias 158
bile 30–31
binary fission 46
binocular microscope 2
bioethanol 139–140
biological sedimentary rocks 89–90
biotechnology 50–54
blind experiments 157–158
blood 35–36
  red blood cells 13, 35–36, 41
  white blood cells 20, 35–36, 41
blood pressure 35
blood vessels 25, 35
bonds, chemical 68–69, 74
bones 41
breathing 36
brittleness 63
bronchi 36
bronchioles 36
Bunsen burners 152–153
burns 152
bushfires 119

C

cambium 26
canine teeth 30
capillaries 25, 35
cars
  driving speed 124
  fuel for 125, 139–140
carbon dioxide 35–36
carbonic acid 84
cardiac muscle 42
carnivores 30
cast 106
cells 2
  in animals 8
  in plants 9, 26
  structure of 6–11
  types of 25
cell division 12–16, 46–47
cell membrane 8–10, 13
cell theory 8
cellular respiration 35–36
celluloid 69
cellulose 31
cell wall 9
cementation 89, 95
changes of state 55–60, 73
charcoal 106
chemical bonds 68–69, 74
chemical formula 68
chemical potential energy 125, 129
chemical reactions 72–76, 129
chemical safety labels 152
chemical sedimentary rocks 89–90
chemical weathering 84
chloroplasts 8
chromosomes 13–14, 46–47
cilia 9
circulatory system 34–38
clast 90
clastic sedimentary rocks 89–90
cleavage 79
clones 46
cloning 51–52
coal combustion 129–130
cold *see* freezing
colour (of minerals) 79
compact fluorescent lamps 139
companion planting 118
compounds 66–71
compression 58
concentration 58
conclusion 157, 164
condensation 58
conduction 63, 134–135
conservation of matter 73
contact metamorphism 94–95
contraception 51
controlled variables 147
convection 135
copper 106
cortical cells 26
country rock 94
covalent bonds 69
crop rotation 118
crossing over 14
crust 78
  metamorphism in 94–95
  ores in 100
  weathering of 82–87
cryolite 107
cryopreservation 52
crystals
  in minerals 78–79, 100–101
  in rocks 94–95
cytoplasm 8

D

data collection 158
data interpretation 157
daughter cells 13–14
decomposition reaction 74
defecation 31, 40
density 57
dependent variables 147
diagrams 162
diaphragm 36
diffusion 8, 58

INDEX

digestive system 29–33
discussion section 164
disease 17–23
  infectious 19–20, 25
  non-infectious 21
  prevention of 21
  transmission of 19
dissection 153
diversification 118
double-blind experiments 158
drugs 19–21
ductility 63
duodenum 31
dynamic metamorphism 95

E

earthquakes 113
Earth structure 77–81 *see also* crust; rocks
ecosystems 112
egg (ovum) 14, 47, 52
elastic potential energy 124, 129
electrical conductivity 63
electrical energy 129–130, 140
electrolysis 106
electrolytic refining 106–107
electromagnetic radiation 129, 135
electrons 62–63
electron microscope 3
elements 61–65
ellipsoidal joint 41
energy
  in body 8, 125, 129
  in chemical reactions 74
  electrical 129–130, 140
  heat 129–130, 133–137
  kinetic 122–127, 129, 134
  potential 122–127, 129
  solar 129, 135, 140
energy conservation, law of 130
energy efficiency 138–143
energy transfer 129
energy transformation 128–132
environmental issues
  energy efficiency 138–143
  health 21
  pollution 110–115
  sustainability 116–121
erosion 82–87
  caused by farming 118
  rocks formed by 89, 95
erythrocytes (red blood cells) 13, 35–36, 41
ethanol 139
evaporation 58
evidence 146
excretion 40
  defecation 31, 40
  respiration 35–36
  urination 40
excretory system 39–44
expansion 58
experiments 146 *see also* investigations
experimental reports 161–166
extrusive igneous rocks 94
eye protection 152

F

faeces 31, 40
families of elements 62–63
farming 118–119
fertilisation 14, 47
fires
  bushfires 119
  heat energy from 135
  in laboratory 152–153
first-hand investigations 156–160
flagella 9
flowering plants *see also* plants
  parts of 25–26, 47
  reproduction in 47
foliation 94
food 30
  energy in 125
  production of 118–119
fossils 88–92
fossil fuels 139
freezing
  of gametes 52
  of liquids 58
  weathering caused by 84–85
friction 130
frost wedging 84
froth flotation 106
fruit 47
fuels 125, 139–140
fungi 9, 19

G

gametes (sex cells) 14, 47, 52
gangue 106
gases 57–58
gas exchange 9, 36
generalisation 146
genetics
  in cell division 13–14, 46–47
  in cloning 51
  in plant breeding 118–119
geological period 90
germ theory of disease 19
glaciers 85, 89
glucose 8, 129
graphs 164
gravitational potential energy 125, 129
groups of elements 62–63
group work 157
guard cells 9

H

habitats 112
hardness 78–79
health triangle 21
heart 35, 42
heat
  measurement of 134
  metals extracted by 106
  rock formed by 94–95, 101
  weathering caused by 84
heat conductivity 63, 134–135
heat energy 129–130, 133–137
herbivores 30–31
high grade ores 100
hinge joint 41
Hooke, Robert 2
hornfels 95
houses, energy consumption in 139
hydration *see* water
hydrogen 62
hydrothermal deposits 101
hypothesis 146

I

ice *see* freezing
ice cracking 84
igneous rocks 93–97
immune system 19–20
immunisation 20
immunosuppressants 52
impressions 90
incisors 30
independent variables 147
infection 19–20, 25
inferences 146

INDEX

insulation 135, 139
intestines 31
intrusive igneous rocks 94
investigations
  experimental reports 161–166
  fair testing 144–149
  first- and second-hand 156–160
  laboratory work 150–155
in-vitro fertilisation (IVF) 51
ionic bonds 69
ionic compounds 74

J

joints 40–41
joules 124, 134

K

karst topography 84
kidneys 35, 40, 52
kilns 106
kilograms 124
kinetic energy 122–127, 129, 134

L

laboratory reports 161–166
laboratory work 150–155
landscape 84
large intestine 31
latent (hidden) heat 134
lava 94, 100
law
  definition of 146
  of energy conservation 130
lead 111
leaves 25
lenses in microscopes 2–3
leucocytes (white blood cells) 20, 35–36, 41
lighting 139
line graphs 164
liquefaction 58
liquids 57–58
lithosphere 78
liver 30–31
low grade ores 100
lungs 36
lustre 78
lymphocytes 20

M

macrophages 13
magma 94, 100–101
magnification 3
magnifying glass 2
malleability 63
mantle 78, 94
mass 57
materials lists 162
matter 55–60
  definition of 57
  conservation of 73
  particle model of 57
  state changes 55–60
measurement, units of 57, 124, 134, 139
meat eaters 30
meiosis 14, 47
membranes, in cells 8–10, 13
meningitis 19
metabolism 40
metals 98–103
  extraction of 104–109
  native 100
  in periodic table 62
  in reactions 73–74
  recycling of 111
metamorphic rocks 93–97
metamorphism 94–95
meteorites 100
metres per second 124
microbes 19
microorganisms 19, 46
micropropagation 51
microscopes 1–5
minerals 77–81
  definition of 78
  in fossils 90
  identification of 78–79
  in ores 100
  in rocks 89, 94
mining 104–109 *see also* ores
mitochondrion 8
mitosis 13–14, 46
Mohs scale 78–79
molars 30
molecular compounds 74
molecules 68
monocular microscope 2
mouth 30–31
multicellular organisms 24–28
muscles 41–42
musculoskeletal system 39–44

N

native metals 100
natural disasters 113
nephrons 40
neutrons 62
non-communicable disease 21
non-infectious disease 21
non-metals 62, 73–74
nuclear fusion 129
nucleus 8

O

observations 146
Oersted, Hans 106
oesophagus 31
omnivores 30
open investigations 157
ores 98–103
  environmental issues 111
  metals extracted from 104–109
ore body 100
organs 25
organelles 8
organ systems 25
organ transplantation 52
ovum (egg) 14, 47, 52
oxidation 84
oxidisers 125
oxygen 30, 35–36

P

palisade cells 9
pancreas 30–31
parent cells 13–14
particle accelerators 62
particle model of matter 57
pathogen 19, 25
pathogenic theory 19
pebbles 90
periodic table 62–63
petrified fossils 90
phagocytes 20
pharynx 36
phloem cells 9, 25–26
photosynthesis 9, 129
photovoltaics 140
physical changes 73
physical weathering 84
pivot joint 40
placer deposits 101
planning of investigations 157
plants
  cells in 9, 26
  cloning 51

INDEX

effect of bushfires on 119
farming 118–119
fossils of 90
parts of 25–26, 47
photosynthesis in 9, 129
reproduction in 46, 47
structure of 25–26
weathering caused by 84
plant eaters 30–31
plasma 35
plastics 69, 113
platelets 35–36
plutonic rocks 94
pollination 47
pollution 84, 112–113
polymers 69
polystyrene 69
porosity 89
potential energy 122–127, 129
power stations 129–130
predictions 146
premolars 30
pressure, rock formed by 90, 94–95
primary source 158
problem solving *see* investigations
procedures, laboratory 163
propagation 51
protein 41
protista 9
protons 62–63
protoplasm 8
protozoa 2, 19
pure substance 62

R

radiant energy 129, 135
reactions, chemical 72–76, 129
re-crystallisation 94–95
recycling 111–112, 118
red blood cells (erythrocytes) 13, 35–36, 41
regeneration 119
regional metamorphism 95
reliability 147
replication 13
reports 161–166
reproduction 14, 45–49
reproductive cloning 52
reproductive technology 51–52
respiration 35–36
respiratory system 34–38
results section 163
rocks
igneous 93–97
metamorphic 93–97
minerals in 78–79
sedimentary 88–92, 95
weathering of 84–85, 89, 95
rock cycle 95
roots (of plants) 25, 84

S

safety, laboratory 152, 162
Schleiden, Matthias 8
Schwann, Theodor 8
scientific method 144–149 *see also* investigations
secondary source 158
second-hand investigations 156–160
sedimentary deposits 85, 89, 101
sedimentary rocks 88–92, 95
selective breeding 118–119
semi-metals 62
sex cells (gametes) 14, 47, 52
sexual reproduction 14, 45–49
single-celled (unicellular) organisms 2, 9, 51
skeletal muscle 42
skeletal system 39–44
skin 25
small intestine 31
smelting 106
smooth muscle 42
solar energy 129, 135, 140
solidification 58, 89
solids 57–58
somatic cell nuclear transfer 52
sound energy 124
speed signs 124
sperm 14, 47, 52
stamen 47
state, changes of 55–60, 73
stems 25
stigma 47
stomach 31
stomata 9
streak 79
styrofoam 69
Sun, energy from 129, 135, 140
sustainability 116–121
symbols for elements 62, 68
synthesis reaction 73–74

T

tables 163
team work 157
teeth 30–31
temperature 134 *see also* freezing; heat
testing 146 *see also* investigations
test tube holders 152
theory 146
therapeutic drugs 21
thermal radiation 135
tissues 25
tissue culture 51
tissue transplantation 52
title of report 162
trace fossils 90
trachea (windpipe) 36
transmission of disease 19
transplantation 52
transportation of sediments 89, 95, 101
tsunamis 113

U

underground mining 106
unicellular (single-celled) organisms 2, 9, 51
urination 40

V

vaccination 20
vacuoles 8–9
van Leeuwenhoek, Antoine 2
vaporisation 58
variables 147
vascular bundles 25–26
vegetative propagation 51
veins 35
velocity 124
vena cava 35
ventilation 36
vertebrates 25
viruses 19
volcanic rocks 94
volume 57

W

water
in chemical reactions 129
erosion caused by 84–85
ores formed by 101
pollution of 112–113

INDEX

recycling of 111–112
weathering caused by 84
weathering 82–87, 89, 95
wetlands 111–113
white blood cells (leucocytes) 20, 35–36, 41
wild cabbage 118–119
wildlife 112
wind energy 129
wind erosion 85
windpipe (trachea) 36
wind pollination 47

X

xylem cells 9, 25–26

Z

zygote 14, 46